ABSOLUTE
MOTION

Rupesh Sharma

ISBN-13:
978-1508519560

ISBN-10:
1508519560

© 2015, Rupesh Sharma
2015 Edition, Revised-Oct 30, 2015

DEDICATED

TO

KISHOR KUMAR SHARMA

(मेरे प्यारे भाई)

TABLE OF CONTENTS

PREFACE

There was a boy, quite innocent, pretty studious, intelligent and far from heaving any bad habit. Somehow or anyhow he got an admission in a college, far away from his home and as it happens, things started changing. He was curious by nature, so, while he was studying "Human Blood Coagulation" as a part of his first year degree syllabus, he saw a possible link between Hematology and Electromagnetism and he was curious to know that what would happen if a "Time Varying Electromagnetic Field" is applied on this natural process, when human blood clots and he even searched the answers, all over the internet and in scientific literature too. Although he got few related stuff, still not exactly what he was looking for and then he decided that he had to do it by himself and he made his own Electromagnets and electric circuits in his study room (Where the money came from? Obviously, from his tuition fee) . But there was one more thing that was needed; where from the human blood will come, in order to perform the experiments? So, he used his own blood and as it is the case with every honest dedication, he finally got impressive results, something never happened in science before and he gave even a demonstration of his experimental results to his institution and the institution was convinced that there was something new. Even the engineering college that helped him to perform his experiments, by providing Hall Effect Kit, etc forwarded an affirmative report of his work to his institution and even the vice chancellor of the university told the principal of his institution to consider what he was saying and the next and final step was just publication of his work but as it happens, it never happened. Partly because his institution realized that because of this, he was getting deviated from the main stream. And because while working on his project, he came across with an idea known in science as "Field" and a curiosity that suppose you take human blood clotting as an example and say there is an intrinsic interaction between an Electromagnetic Field (EMF) and human blood clotting. Now, the obvious question is "What the hell it is that is interacting? And if it is interacting with the human blood, it could interact with the whole human existence too".

The question "What is field?" was the curiosity that was enough for the destruction of his quite promising career outright. Because even he was not interested in the project, he was working upon anymore. Now, he had seen an immense potential inherent in the idea of "Field". And in order to find out the answers of his curiosity, he decided to study everything, whatsoever was linked with the subject. But unfortunately at very first place, he didn't realize that whole science is basically linked to the subject, he was curious about. There was one more point connected to his curiosity about the existence of a field and it was the very fact that the idea of field is intimately linked to one of the absolute human curiosity, which is still mysterious that asks "Where I am?" or "Why I am in this apparently infinite universe?" That is also related to other unresolved questions of humanity like "Where we came from?"

Well, he died in very young age but the only question I have in my mind is that "Did he have the answers of his curiosity?"

WHITE BOOK

[A]

So, please sit. Make yourself comfortable. This may take a while. Can I get you a coffee? You got it.

Okay. How do I want to do this? They have made it as complicated as they could. So, it's hard to just dive in. You know what; it is as easy as anything but perhaps they tried to make it hard, so that the full control remains only in the hands of a bunch of people, what they call "Specialists". So, if you feel like I am off to the side of the tale half the time, well, this is why. Just bear with me, and we will get to the end in good time. Okay?

Okay, let's see.

[B]

Imagine, you wake up one day and your room is gone, your bed too is gone. Hell, everything is gone and you discover that you are in an inky void, where except you nothing is present, not even a star, like an empty space that lies between the Sun and the Earth. Well, you might say, it's a dump idea but just go with it, I assure you, it is not. And you as well as I can see that it is not really hard for you to imagine that one day you wake up in a perfectly empty void. If you could wake up in your dream world, you could wake up in this inky void too. No big deal. Or is it?

Okay, so, let's say again, one day you wake up in an inky void, where you are the only existence (Make a note of that). Your existence in this void might appear like this.

[As a matter of fact, the appearance that you are the only existence in this inky void might be elusive because there is one more thing with you here and it is this darkness, all around you. But it is not something, you would like to consider. Why not? Perhaps because your human senses can't tell you, what it is. So, it seems nothing to you. Now, the question is, is it really "Nothing"? Let's see.]

[Note: In order to achieve the best simplification possible in this book, I have a proposal for you. I am giving you a human body or a geometrical object that looks like this.

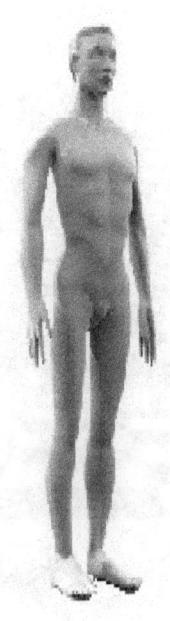

And what all you need to do is that just place yourself or your soul in this body and follow this geometry, wherever it goes and then this whole journey of ours will be easy as well as perfect.]

Now, you start exploring your present domain that is this mysterious void and after some time you realize that there is no way to find out, how big or small this void is and where it is in the universe. But soon you realize that as you are standing still, there must be a surface beneath your feet. So, you are standing alone on the floor of this inky void. It would be convenient, if we draw a picture of your location in this void and here it is.

Suddenly, you feel a jerk, a sort of force on your body that expels you forward. Analogous to a force you experience, when you board a metro train and suddenly the train stops at a platform and you feel a force on you. Because you have your teachings with you, so, you conclude that there must be a change in the state of motion of this inky void, as per the concept of "Inertia" and Newtonian or Classical Mechanics, as you can see right here.

THE

MATHEMATICAL PRINCIPLES

OF

NATURAL PHILOSOPHY.

DEFINITIONS.

DEFINITION I.

The quantity of matter is the measure of the same, arising from its density and bulk conjunctly.

THUS air of a double density, in a double space, is quadruple in quantity; in a triple space, sextuple in quantity. The same thing is to be understood of snow, and fine dust or powders, that are condensed by compression or liquefaction; and of all bodies that are by any causes whatever differently condensed. I have no regard in this place to a medium, if any such there is, that freely pervades the interstices between the parts of bodies. It is this quantity that I mean hereafter everywhere under the name of body or mass. And the same is known by the weight of each body; for it is proportional to the weight, as I have found by experiments on pendulums, very accurately made, which shall be shewn hereafter.

DEFINITION II.

The quantity of motion is the measure of the same, arising from the velocity and quantity of matter conjunctly.

The motion of the whole is the sum of the motions of all the parts; and therefore in a body double in quantity, with equal velocity, the motion is double; with twice the velocity, it is quadruple.

DEFINITION III.

The vis insita, or innate force of matter, is a power of resisting, by which every body, as much as in it lies, endeavours to persevere in its present state, whether it be of rest, or of moving uniformly forward in a right line.

This force is ever proportional to the body whose force it is; and differs nothing from the inactivity of the mass, but in our manner of conceiving

["Mathematical Principles of Natural Philosophy" by Sir Isaac Newton-
Published by Daniel Adee, 45 Liberty Street, New York]

13

[Look, in the explanation of "DEFINITION 1", he is saying,
"..............I have no regard in this place to a medium, if any such
there is, that freely pervades the interstices between the parts
of the bodies.........."

You know what, with all due respect, Izz was completely off.
As we are about to see, it is this goddamn medium itself, who is
the very reason for the existence of the same thing, he is trying
to define, "The Mass".]

As Izz (Isaac Newton, 1642-1726, England) said,

"The vis insita, or innate force of matter, is a power of resisting by which everybody, as much as in it lies, endeavors to preserve its present state, whether it be of rest or of moving uniformly forward in a straight line."

[Page-73, Philosophiae Naturalis Principia Mathematica, 1687]

[Please make a note of this, "A power of resisting" because Newtonians think that this power exists within the body itself and that is one of our main concerns in this book, as we need to discover that where this power is coming from. Because it appears that an inert can never resist or endeavor by any means.]

And as we have taught in our schools, by our teachers that in the simplest way, it means,

"In an inertial reference frame, an object either remains at rest or continues to move at a constant velocity, unless acted upon by an external force."

So, you conclude that as I am the only body in this inky void and I experience a force that pushes my body in a certain direction (What we say forward direction because you can see that there is no way in this void that you can define the concept of dimension or direction). So, there must be a change in the state of motion of this void (We said state of motion because you certainly have no way to know here that whether this inky void was in motion or on at rest before you boarded it. Worse still, just as a consequence, even now you don't know, whether the void that is your present domain is in motion or on at rest). Okay, now the question is how you arrive on this conclusion? And here begins a very interesting story, story of the birth of what we call today as "Modern Science" and its evolution in more than 3000 years (or even older than that).

Date back to the establishment of a difference between what is scientific and what is non-scientific, it was Ari (Aristotle, 384-322 BCE, Greece) who by experimentation and logic, convinced his people

"The natural state of a physical object is 'at rest' and if it sets in motion there must be a force to act on it."

Simple though, this conclusion remained as a central dogma for years, as our human common sense even today says the same thing, until Gal (Galileo, 1564-1642, Italy) came in and tried to see whether you really need a force, if a natural object is in motion and like Ari, Gal also performed number of interesting experiments and finally concluded that

"No, Ari, if an object is in motion with a uniform velocity, you don't need to give it a push."

And he arrived on even a stranger conclusion that in contrast to Aristotelian physics, the state of at rest and state of uniform rectilinear (in a straight line) motion are one and same things. They are just two faces of a single coin. They are two different manifestations of a single reality. Simple it might appear to you but this was one of the point that gave birth to a revolution known today as "Copernican Revolution", a perfect paradigm shift in the history of human civilization that brought our Sun at the centre of this universe instead of Earth, as it used to be the general belief of that time (And it even made Gal "Father of Modern Physics") and a turning point when what used to be called as "Natural Philosophy" became what we call today as "Modern Science". Well, happened, happened anyway. Why I am making it an issue now? Because I realize that we need to reinvestigate this subject because the concept of inertia as it is taught throughout the globe is one of the top notch fundamentals of modern physics that is always linked to other sciences as well. So, if there are really some discrepancies in it, it will lead to disastrous consequences and because it seems that something that has an intrinsic value in this idea, is still missing.

Well, before we go ahead, we need to realize that "Is it really important to investigate an idea, which is now a part of physics text books? And the best possible answer I can come up with is "Yes", as even though this simple happening that happens with you, me or with everyone, here on Earth, is not as simple as it appears.

"Is it possible that this everyday human experience on Earth might have a cosmic connection? Or, it is just a local phenomenon?"

Okay, let's say that we need to investigate the subject in a best possible scientific way and let's say that there is a physicist inside of you. So, here we go.

You are present in an inky void, whose extent of expansion you don't know and you also don't know whether this inky void that is your present domain is in motion or it is on at rest somewhere in the universe, as you don't have any way to look outside of it (Look, if you can realize, same is true for your real void, sorry, your real world too but just in a bit different context). What you all know is that, you have a floor on which you are standing still. Suddenly, you feel a jerk on your body, a force that expels you in a certain direction. Now see, before this event, there was no way for you to define the idea of dimensions or direction. You simply could not define, what do you mean by 'Forward' or 'Backward' or 'Left' or 'Right'. But as you can see, after this happening, you can say that the force that expelled you in a certain direction, implies that it is a forward direction and its opposite direction is backward direction and you can even define the left and right directions as well (All in relation to you). Even better you could define what you mean by 'East' and 'West', if I give you the Sun back. So, the situation now appears something like this.

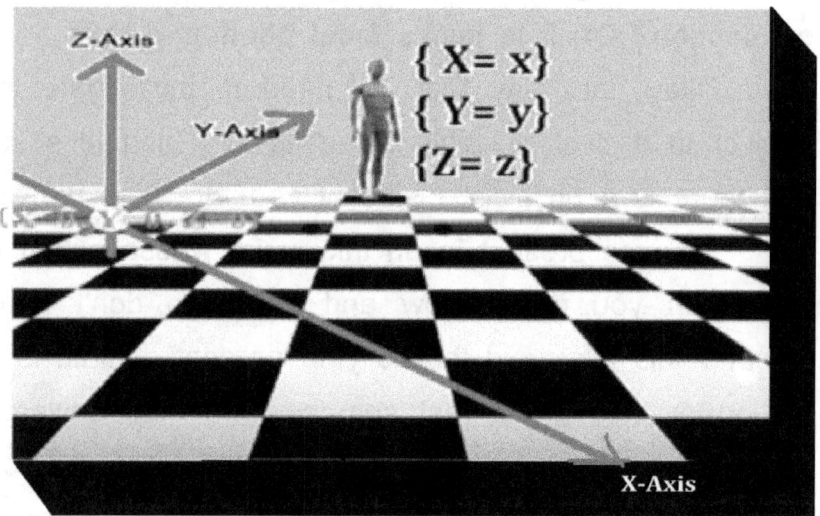

Now, you can define what you say 'Up' and what you say 'Down' too and let's say again (as it is important to realize) that out of a sudden, you feel a force on your body that expels you in a certain direction, which we just defined as forward direction.

Okay, now the best possible scientific explanation you can come up with is this.

Because the body that is the human body I have, possesses mass, so, it must show inertia as "a tendency of an object to resist any change in its state of motion" and as there must be a kind of change in the state of motion of this void, whether it was on at rest or was already in motion before I boarded it that by the law of inertia or Newton's first law of motion, exerts a force on my body and as a consequence I feel a force on myself, as if I am in a bus and bus suddenly stops and I fell down in a forward direction.

Well, this is exactly what is taught to every science student throughout the globe today and as a consequence, we realize that we know what this inherent tendency of a physical object or any physical for that matter is and we call it "Inertia" and also due to the fact that you have a confidence on your Newtonian giants like Ari and Gal and all others including Izz and A1 (Albert Einstein, 1875-1955, Germany).

As a matter of fact, I too used to have it before I entered in this Inky Void. But now, everything looks different.

[C]

[Part 1]

We have already seen that idea of inertia and its science today, as it is explained in classical mechanics is one of the most crucial points in the history of evolution of modern science. Because it was the starting point for Izz to develop, what we today call Newtonian or Classical Mechanics. Even it helped our A1 to formulate his revolutionary "Spatial Theory of Relativity" in 1905 and his "General Theory of Relativity" in 1916. So, if there is even a slightest error in the idea, it will have disastrous consequences for the whole body of modern science. A chaos? Indeed.

There is something to which I would like to call it as "IDS" that is "Information Deficiency Syndrome". Ari, Gal and all others including Izz and A1, really had no way to know, what they were dealing with. Because with the advent of technology, we have a huge amount of information now, in this dawn of 21st century human civilization on the planet Earth, about the creation and the universe as a whole. So, the point is that, a misunderstood beginning must end up on a misunderstood end. And this is exactly what we have today on the name of modern science on this subject and now, I show you how this idea of IDS (Information Deficiency Syndrome) works. We reload the same experiments that were used earlier, to define the subject and to explain the concept of inertia but in a little different way.

As you can see, in order to work out the truth of a natural phenomenon that happens on Earth surface, it is equally important to find out, your true position as an observer of that same phenomenon, in the same domain that is this universe. Because you and the phenomenon you observe are not really isolated from each other, just because you both exist within a common domain that is this creation. Because, the working of this universe is such that sometimes as the situation gets changed, the truth also gets changed with it. Now, say you are in a metro train and the train is travelling between one station to another station, after some time when the station comes, train stops and you feel a jerk, a force that expels you in a forward direction and if you ask anyone why is it so, undoubtedly, you are given an explanation that as your physical body possesses mass and hence it shows inertial nature. So that when the train stops, there is a change in your state of motion, from a moving phase to at rest and it is this change that gets conveyed by the floor of the metro train, which exerts a force to expel you forward. Perfect, as I said there is IDS (Information Deficiency Syndrome), so now, we begin to reinvestigate this scientific explanation in a little depth.

No offence but neither Ari nor Gal or any other genius for that matter had any idea that the phenomenon they were looking at, doesn't only have a local aspect (Even Earnst Mach who realized that something is not right also had a quite similar idea to which I consider incomplete.) but as we have guessed that this simple human everyday experience here on Earth might have a perfect cosmic connection. And as discussed, we really need to discover the ultimate source of this "Power of Resistance" that belongs to every single physical existence. So, in search of this cosmic power, we are about to depart from our present vantage point on earth surface, towards the horizons of our known universe with a very high speed like a supersonic or even faster than anything we can imagine. Well, as you know, modern physics defines an upper limit to everything that can move in this cosmos or in this creation. They say it is velocity of light, "c" that is equal to 29, 97, 92,458 meters per second (or say it is just 3×10^8 m/s). Even they have developed a science to prove this assumption and that's the reason, it's not an assumption anymore but a natural fact for us. But our time is limited, so, right or wrong, let it all go to hell, we need to travel faster than this and we will for sure.

So, somehow, someway, we acquire a great velocity such as "Velocity of Human conscience" or like "velocity of human imagination" (or "Man ki Gati?" may be). We will use multiples of powers of tens, like ten to the power one (10^1), ten to the power 2 (10^2), 3, 5, 10, 20, and 40 and so on, in order to measure cosmic distances. And what all you need to do is just forget about everything else, like your family, a beautiful girl sitting next to you in the metro. Even you need to forget your own physical existence for a while because at this point of time I feel that it is quite probable that the true and absolute domain of all your limits is this geometrical object, what you call your physical or human body. Never mind. We need to take off now, with this imaginary velocity, towards the edge of this universe, in order to find out the ultimate truth. And in order to start, let me presume that one day you board a metro and travelling between two metro stations, say New Delhi to Gurgaon. So, you have plenty of time to think about anything you like. Well, of course you can also enjoy music, if you want to but let's say you prefer thinking. [Just to make it easier, we can draw a tour plan for us that might look like this.]

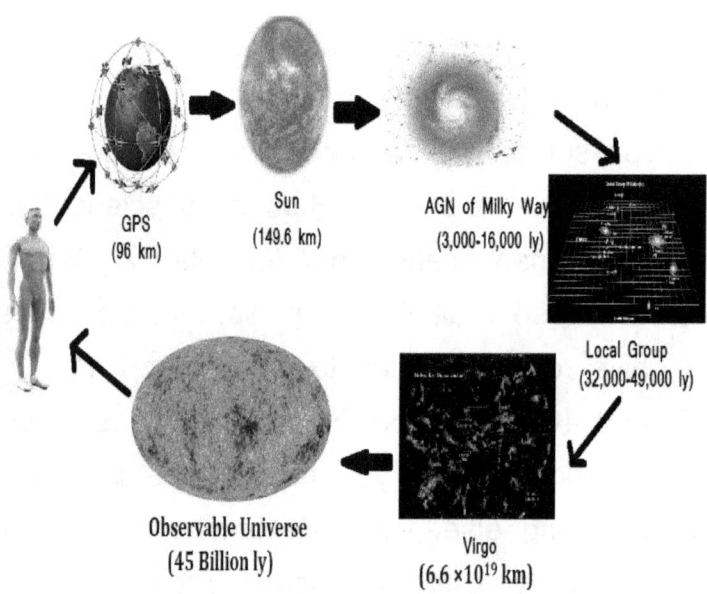

GPS
(96 km)

Sun
(149.6 km)

AGN of Milky Way
(3,000-16,000 ly)

Local Group
(32,000-49,000 ly)

Observable Universe
(45 Billion ly)

Virgo
$(6.6 \times 10^{19} \text{ km})$

[This is only half of the tour plan and we will get the next half, when the time comes.]

So, here comes a point, we can begin with. Our world, worm, comfortable, familiar but when we look up, we wonder, do we occupy a special place in the cosmos or not? In modern science there is something that is relevant to this question and it's called "Principle of Terrestrial Mediocrity" or "The Mediocrity principle" and it suggests that there is nothing very unusual about the presence of mankind on Earth. Simpler would be that you don't need to think that your existence in this creation or for this creation is special. But a new question is that our one and only domain that is this universe, ever really wanted that we supposed to be here, or we supposed to host it? So, now in this metro train, there are only two options you have. Either you can 'wire in' or you can leave home for the ultimate adventure. To discover wonders, confront horrors, beautiful new worlds, dark forces and even the beginning of time, the moment of creation itself. Because modern science says that the concept of motion through space in this universe is intrinsically linked to our idea of time, what A1 figured it out in terms of "Space-Time". Here is only one way to find out all of it and that would be your own decision. Let's say your decision is positive and you decide to leave your home and your human body, right here in the metro and we start this adventurous journey, presuming that we will be back, before our metro station Gurgaon arrives that would be far better than standing on Earth, looking up in the sky and wondering, what the hell it is. How we can travel deep into space, without a space ship? That's the real issue. But the problem is that "Science of Dimension" or simply concept of dimension, as it is written in the physics text books and as a consequence, it is also what you have in your mind is far from the truth and idea of dimension is as mysterious as it used to be 3000 years ago but because we need to focus on our main theme. So, as of for now, we say that there is one more dimension beyond 3D space or 4D space-time that exists not outside but within the human body itself, inside our brain, starting from a point exactly at the center of our forehead and just above the human eyes (Well, the next probability is human heart, but it would be too spiritual.) and if this is the case, it simply means that our fundamental limit to travel through space-time is our own goddamn realization that we need a vehicle for it. So, now nothing is there to stop us, from reaching to the edge of the universe. And 1, 2, 3 go. Your existence at this moment is in this metro that is running on earth surface and it appears like this.

Plant   You are here

Now,

10^0 m (Altitude) = 1 m.

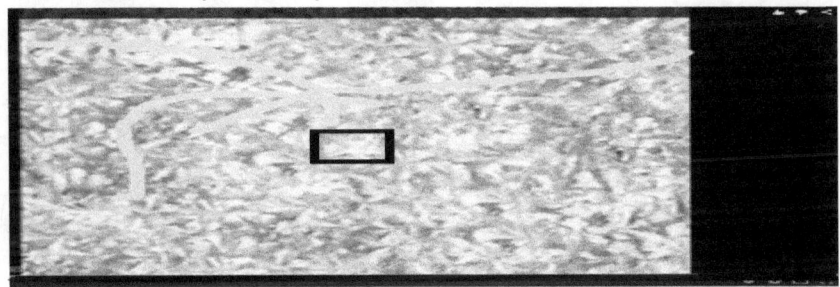

We are 1m above from a plant in the metro, right next to you on the floor. Now,

10^1 m (Altitude) = 10 m.

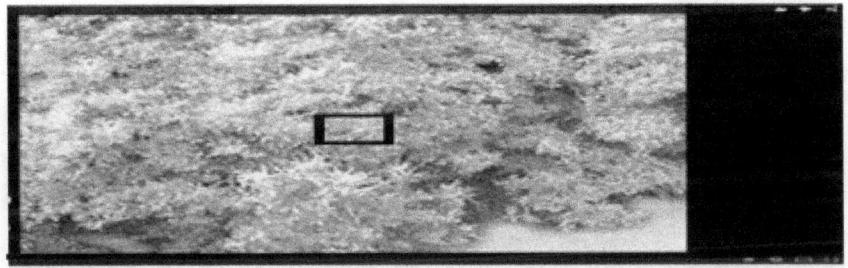

We have started our journey and just left the earth surface. Here we see some trees of the forest in which our metro is passing by. Now,

10^2 m (Altitude) = 100 m.

We can see the edge of a forest as well as metro station itself. Now,

$$10^3 \text{ m (Altitude)} = 1 \text{ km.}$$

We entered the stage of "kilometers" (km). From here you can have a parachute jump, if you like just in case if you are still not ready to leave your sweet home. Now,

$$10^4 \text{ m (Altitude)} = 10 \text{ km.}$$

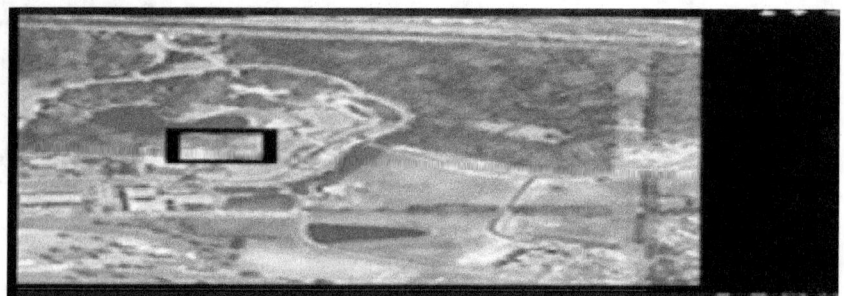

We can see the city and houses with difficulty. Now,

$$10^5 \text{ m (Altitude)} = 100 \text{ km.}$$

You are here somewhere

We see our neighborhood. And now,

10^6 m (Altitude) = 1,000 km.

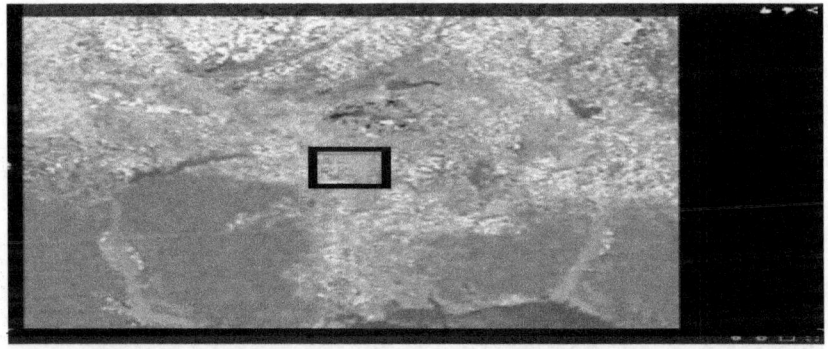

This is a satellite view. Now, we are out of earth atmosphere and above the clouds. See, our journey through time and space logically begins with this single step. At the edge of our space, only 96 km up, where we place our satellites in the orbits around the Earth. Just an hour drive from home. Down there, life continues. Your metro is running on track, everything is looking as if you got a satellite or GPS (Global Positioning System) itself installed inside of you (or you are installed in GPS?). When we return home (if we return home?), will it be the same? Will we be the same? Only future can tell. Now,

10^7 m = 10,000 km.

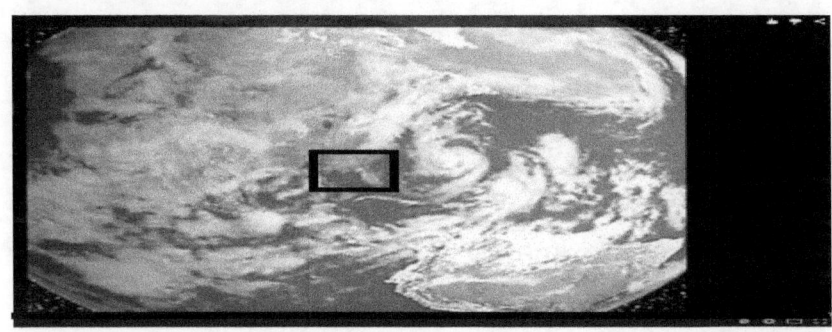

This is the half part of our globe. At this point you can't distinguish your country from her neighbors. Now,

$$10^8 \text{ m} = 1, 00,000 \text{ km}.$$

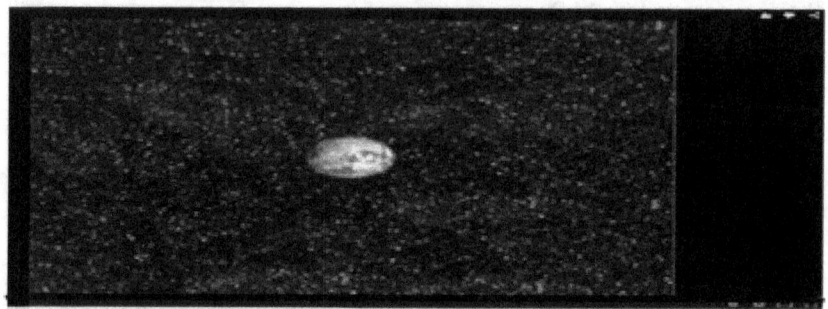

From here we see that the Earth looks like a small ball, surrounded by an infinite darkness (Perhaps the same darkness as we saw when you were in your inky void). Now,

$$10^9 \text{ m} = 1 \text{ million km}.$$

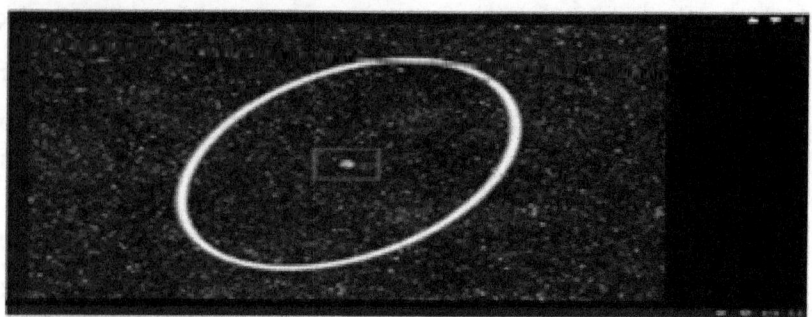

We see Earth and Moon's orbit in the solar system.

We have to leave all this behind to reach Moon, our next cosmic station. Dozens of astronauts have come before us. Just a 3,84,400 km from home and it would take approximately 3 days for them even if they got the fastest space craft available today.

Our Moon, it's like a desert. So close. Neil Armstrong's foot prints look the same as they were formed when Neil himself touched the Moon on 20th of July, 1969. There is no air to change them. They could survive for millions of years, may be longer than us. Our time is limited. We need to take our own giant leap and here we go.

10^{10} m = 10 million kilometers.

Part of the Earth's orbit around the Sun and full orbit of the Moon around the Earth is visible.

Now,

10^{11} m = 100 million km.

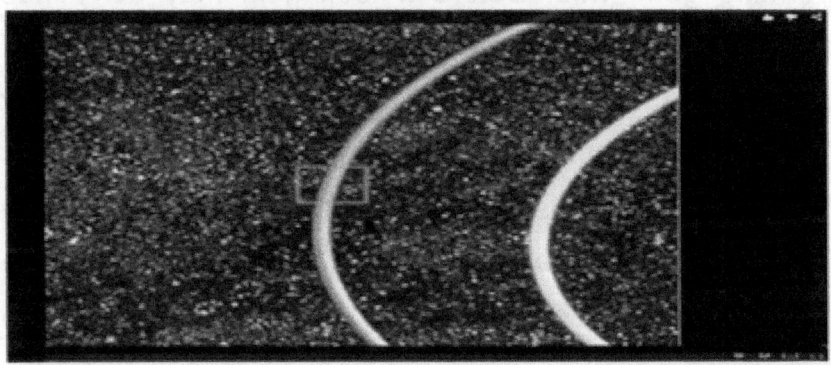

Orbits of Venus and Earth around the Sun are observable to us.

Now, 1 million km, 2 million, 5 million, 10 million, 20 million km. We are far beyond where any human ever ventured. In our approach towards the next cosmic station, the Sun. Out of the darkness, here comes, the goddess of love "Venus", the morning star, the evening star. A sister to our Earth, she is about the same size and gravity as Earth. The Venus Space Probe has revealed that, these dazzling clouds out here, they are made up of sulfuric acid; the atmosphere is choking with carbon dioxide. Venus is one angry goddess indeed. The air is noxious, the pressure is unbearable and it's hot, approaching 900^0C. Nothing can survive here, not even a space probe. So lovely from Earth, up close, this goddess is hideous. There are thousands of volcanoes. All the carbon dioxide is trapping the Sun's heat. Venus is like a cosmic rock that is burning up as if it's global worming gone wild. Before it took hold may be Venus was beautiful, calm more like her sister planet Earth. If global worming remains continuous in our nearby future, probably this could be the Earth's future.

Wait, there is something else, obscured by the Sun. It must be Mercury. This is what happens, when anything gets too close to the Sun. Temperature swings wildly here. At night it is -275⁰C, come midday, it is 800⁰ plus. Burnt then frozen. The Messenger Probe (2004-2015) told us something strange, for its size, Mercury has powerful gravitational pull. It's a huge ball of iron, covered with a thin layer of rock. Now,

10^{12} m = 1 billion km.

Orbits of the planets Mercury, Venus, Earth, Mars and Jupiter are visible.

And, now

10^{13} m = 10 billion km.

From this distance, we see the solar system itself and paths of all the planets in it.

Here is our Sun in its entire mesmerizing splendor. Our light, our lives, everything we do is in his control. Everyone depends on it. It's the Greek god "Helious", driving his chariot across the sky. It's the Egyptian God "Ra", reborn every day. Sun is the face of God (And on top of everything, "Surya saakshhat brahmm hein. Mein surya ki brahmm roop se upasana karta hoon."). It's so far away; if it burned out, we wouldn't know for about eight minutes because light would take approximately 8 minutes, travelling between the Sun and the Earth, to inform us that our Sun is gone. It's so big; you could fit one million earths inside of it. Now, up close, it is unrecognizable. It's a turbulent sea of fire. The thermometer pushes 10 thousand degrees. Hot enough to transform millions of tons of matter into energy every second to keep it shining. We are 5 trillion km from home but it is just a baby step, ahead trillions of km, billions of stars. It's time to start not looking back but looking ahead to step out in a big and wide universe. Now,

10^{14} m = 100 billion km.

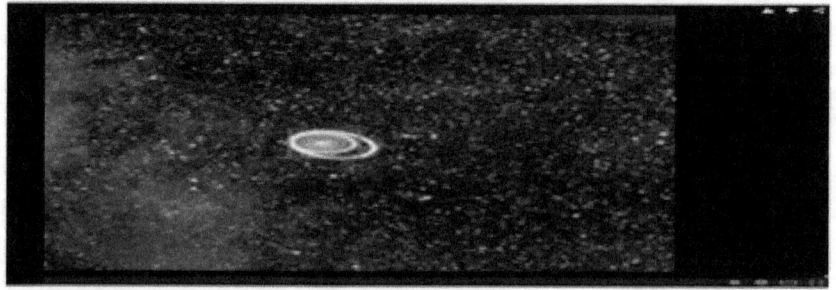

From this altitude, the solar system began to seem small. Now,

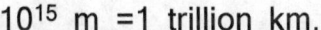

$$10^{15} \text{ m} = 1 \text{ trillion km.}$$

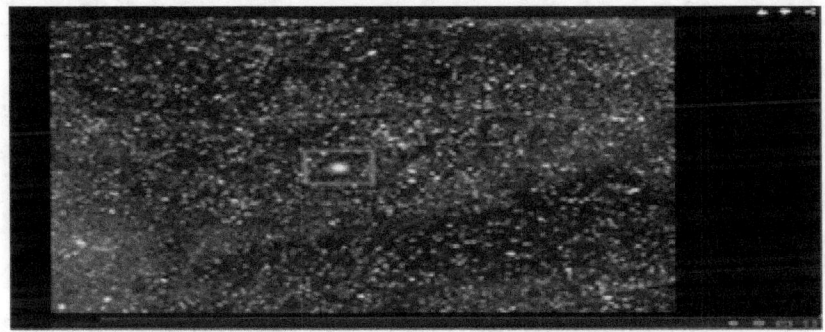

The Sun looks like any other star in the sky, in the mist of thousands of stars.

We are in the "Interstellar space". There are billions of stars like our own Sun, many with planets, and many with moons. It's hard to know, which way to go. There are infinite possibilities. So far, the word "kilometer" is becoming meaningless. Out here we measure our distances from our home on Earth in light years (One light year is equal to 9.4605284×10^{15} meters). Now,

$$10^{16} \text{ m} = 1 \text{ light year.}$$

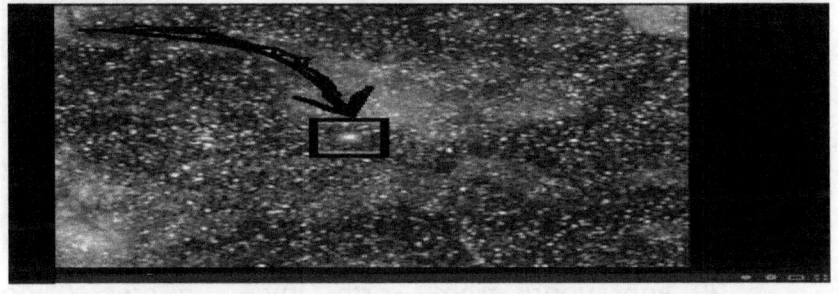

This is our Sun, in our galaxy "Milky Way". Now,

10^{17} m = 10 light years.

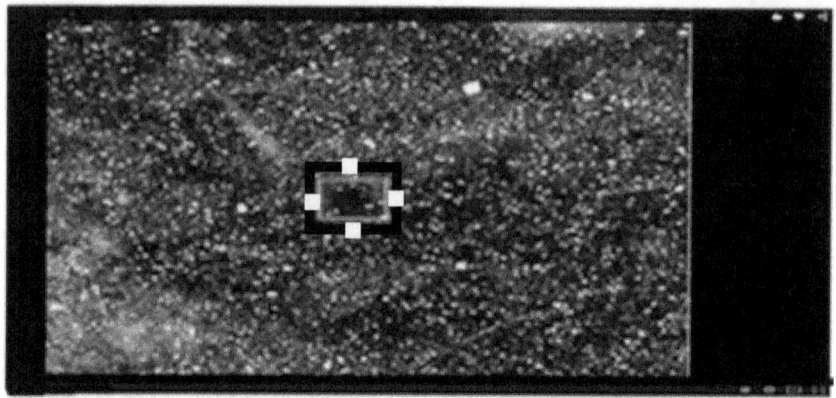

The solar system can't be seen. Now, we are 25 trillion km away from home. A 1, 50,000 years ride in a fastest available space ship and we have only reached just the first solar system beyond the Sun, "Alpha Centauri". We are over 4 light years from home. Distances so vast, they are the mind boggling. Who knows what strange forces lie ahead but hold on it is for certain that we will discover it all when and if we reach the edge of the universe. Now,

10^{18} m = 100 light years.

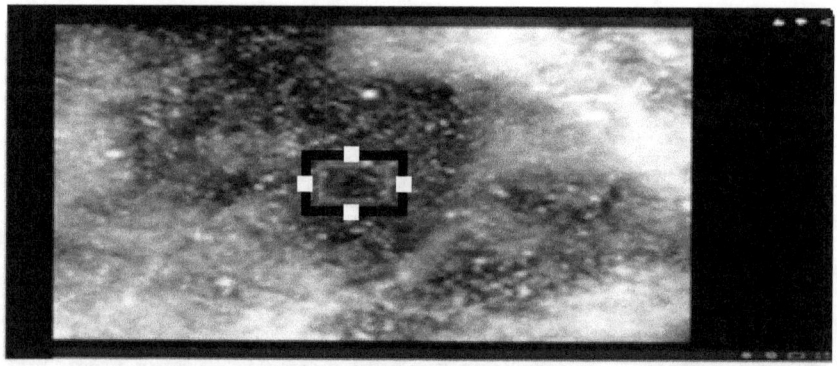

20 light years from Earth, this is a star "Glise

581". Here we find a closest habitable solar system, like our own. In this a planet is at just the right distance from its Sun. And there are ideal conditions for life to emerge and if a comet is ready to deliver water and organic materials, than life, even complex beings like us or even civilizations like our own, could be down there now.

But until we devise a way of communicating over these vast distances, all we can do is speculate. We and they, living parallel lives, unaware of each other's existence or maybe they already know everything about us but just not interested either in Earth or in mankind or maybe they are waiting, waiting for what? Perhaps for a right time? Who knows, if life has come and gone?

Now,

$$10^{19} \text{ m} = 1,000 \text{ light years.}$$

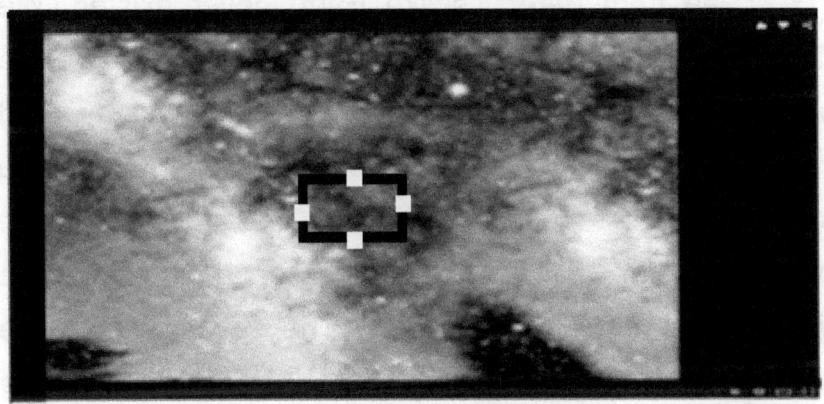

Here we see boundaries of our galaxy, the Milky Way. Keep going, sixty five light years from Earth.

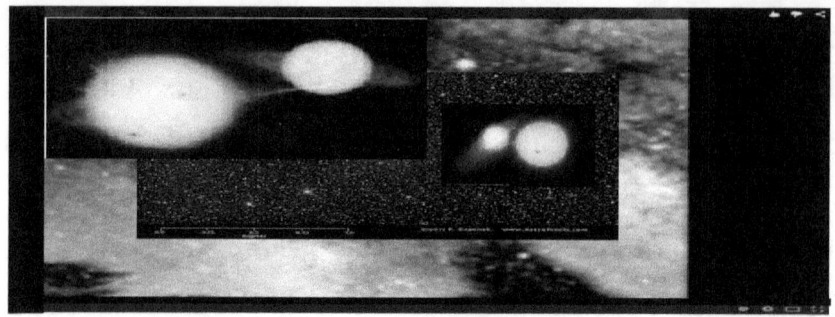

The twin stars of "Algol" and known to the ancients as "Demon Star", up close its even stranger, a perfect demon indeed. One star is being sucked by the other. We are almost 100 light years away from home. Now,

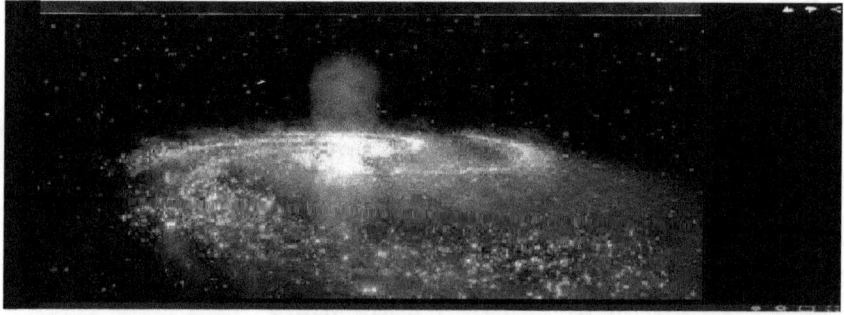

See from here, it's like Earth never existed. We can't find out, where our Sun gone. Looking up at the sky, wondering where and how we fit in. We learned one thing for sure, the universe is too bizarre and too starling for us to guess, what lies ahead. Deep inside our galaxy, the "Milky Way", pinpricks of light that have inspired our old and new tales. The seven sisters, the daughters of the Greek God "Atlas" are transforming into stars to comfort their father, as he held the heaven on his shoulders. Now comes this.

But this, it's not like a star, not a planet, not anything we've seen, a ghostly specter, more than 1,300 light years from Earth, it is "Orion", dark clouds, dust and gas shrouding us. Now,

Their deep inside, light pulling the dust and gas towards itself, heating up, merging into a ball of burning hot gas. Like our star, our Sun in a miniature. Inside its millions of degrees, so hot, it's beginning to trigger nuclear reactions, the kind that keep our Sun shining, making energy and light. A star is being born. We are witnessing the birth of our future universe. We've come to expect destruction but this is one of the universe's greatest acts of creation, the "Star Birth", for a future "Star War"? There are jets of gas, exploding out with tremendous force, blasting dust and gas for millions of kilometers. It's unbelievably violent and creative, "Nebula", vast glowing clouds of gas hanging in space. They seem to be forming a vast stellar sculpture. Nature, it's an artist on the grandest of scales. Now,

$$10^{20} \text{ m} = 10,000 \text{ light years.}$$

Our trip continues in our galaxy. Stars are born, grow up. Do they die? Well, of course, laws of nature if applicable for humans on Earth, exactly in the same way they are true for stars as well. Somewhere between here and the edge of the universe lies the true answer. We observe luminous clouds, suspended in space. Encircling what was once a star like our own Sun. All that is left is these brightly colored gases, elements formed by the nuclear reactions, deep inside and released into space on its death. Green and violet gases are hydrogen and helium, the raw material of the universe. Red and blue are Nitrogen and oxygen, the building blocks of life on Earth or anywhere else. For us to live, stars like this have to die. Every atom in our body was produced by these stellar nuclear fusion reactions. We are made up of all the star stuff and this truly makes us special, no? We are made up of star dust? And now,

Our family tree begins here. At its heart, it is a ghost of a star, a "White Dwarf". It is white, hot, small, but unbelievably dense. As per the science of "Gravitational Collapse", in the stars dying moment, its atoms fuse together, making it so dense that just a teaspoon of this star would weigh approximately 1 ton. Probably, it's a chilling premonition of our Sun's fate. Six billion years from now, our Sun will itself become a white dwarf. Its death will herald the end of life on Earth? Bullshit. There is no way; mankind can make it till then. Makes you wonder, how many worlds have come and gone. Celestial stories left untold, lost forever. But the greatest story of them all is still to be told. We must go back through time to learn how the universe begun. Now,

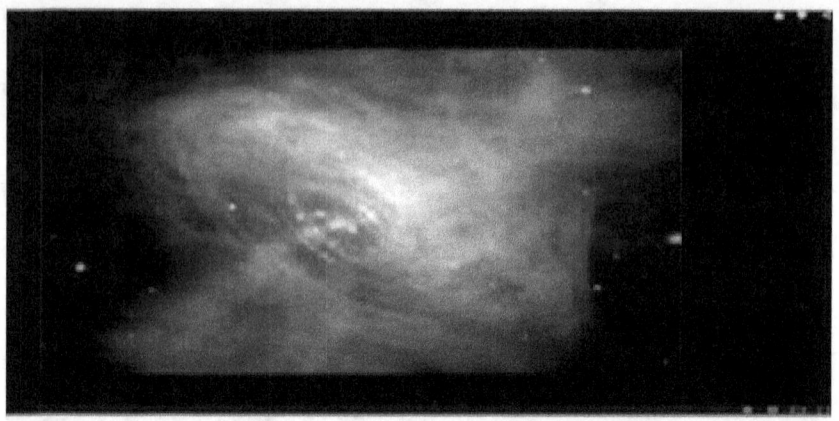

This is "Crab nebula", six thousand light years from home, deep inside a stellar graveyard. We've learned so much, seen things we'd never believed possible. Now, sites like this wonder us beyond imagination, we take our stride. We are ready to face whatever lies behind, determined to reach the edge of the universe.

We are getting a strange sensation, a feeling that there's something bad out here. There is a malevolent presence. One thing we didn't want to encounter. Impossibly black, blotting out the stars behind it. We are staring into the face of extinction, the remains of a giant star, a "Black Hole". Its gravity is so intense that not even light can escape from its grip. Inside, there is no matter as we know it. No time, no space. All the laws of physic collapse on it. Everything whatever was in its vicinity has gone inside of it. Nobody knows where. This is the edge of human understanding.

Perhaps the most important and fascinating thing with the black holes is that it gave us the concept of "Singularity" and in my opinion it is the most important, yet unresolved conundrum of the modern physics.

Like this star, spiraling, disappearing down, an invisible sinkhole, or a great river of fire, might be like this.

Who's to say, we don't live in a vast black hole, the whole universe isn't inside a black hole now or inside another universe? Or we are in a "Warm Hole"? Something like this.

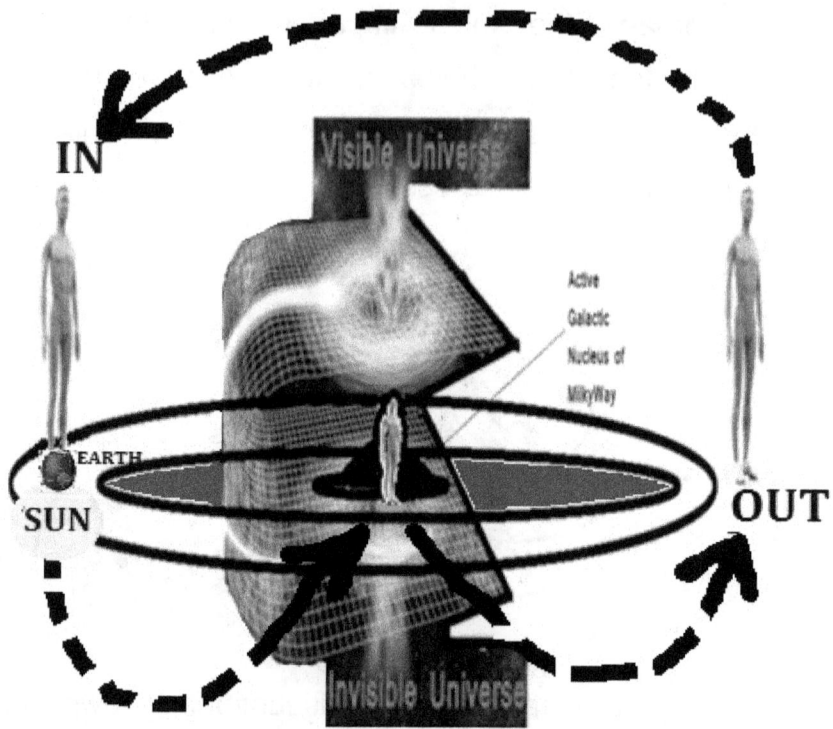

There is or there was a debate on going that if an astronaut crosses the point of safe return or "Event Horizon" (Just in case if you don't know. Event horizon is like a hypothetical covering of a black hole) and keeps on falling into the black hole, what would be his perception about the most fundamental stuff of physics like mass and space and even time. Because it has been speculated and is a well accepted theory of today that time gets stopped itself inside of this black hole singularity. Think about it too long and your mind will real. And we are still in our own galaxy, the Milky Way. Now,

$$10^{21} \text{ m} = 10 \text{ million light years.}$$

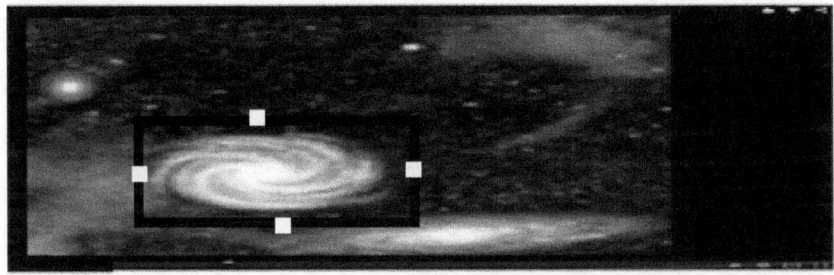

From this far, we see Milky Way as well as her family of other galaxies like M13 or Andromeda.

The vastness of the universe beyond still lies ahead. The wonders, the dangers, the secrets, they're out there, waiting to be discovered. Now,

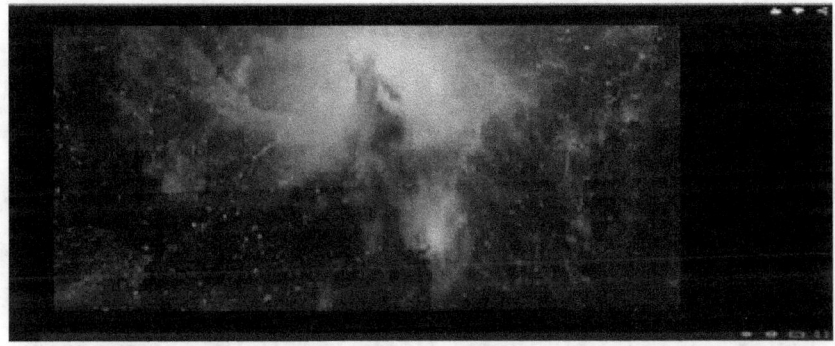

7000 light years from home, it's though we are in a forest. So beautiful, so fascinating, it's impossible to look beyond, to see the bigger picture. We have to find a way through to reach at the galaxy's edge. But faced with sites like this, it's hard to leave. A colossal glowing cloud topped by great amounts of dust, the pillars of the creation. Like a gateway into the unknown. Dazzled by the Milky Way's beauty, we've been blinded to its terrors and strayed into a cosmic minefield like an explosion in slow motion. Our home galaxy, the Milky Way, we wanted to know where we fit in. Here's our answer. Civilizations, past and present, everyone that ever lived, the smallest creäture, the greatest mountain, all of it has gone invisible, not even a tiny speck. It might appear that our home is a minor planet, orbiting around an insignificant star. If it gets disappeared, who would even notice? So, do we now have our previous question's answer? Universe is not even interested in our existence, in this creation or does it? And yet, so far, we've found that nowhere else we could rather live. It's only now, far from home that we're beginning to truly appreciate it. Look at all these stars, the hundreds of thousands of them. Surely one of them, more than one, must be capable of supporting life, may be here in this swarm of stars, the "Great Cluster". I tell you what. Back in 1970s, astronomers sent a message in this direction, detailing our DNA structure, solar system's location, population of earth and some other useful facts about us, if there was a listener or simply put an "Alien species". But the message won't arrive here for another 25,000 years. Regardless of the issues related with the expenses on such sci-fi experiments that are a real global concern today, it at least gave Hollywood a blockbuster "Species" and the best of this whole was Natasha.

No? We haven't found alien life yet but neither have we found any reason to believe, it isn't out there somewhere. There is an equation devised to estimate the number of other advanced civilizations. The result is startling. Mathematical calculations predict that there could be millions of other civilizations, just as our own, within this Milky Way. Approximately everything we've seen so far is inside the Milky Way. Now we're ready to leave our home galaxy, to enter into "Intergalactic Space". Here is our chance to solve the ultimate mystery and to experience the moment of creation. Beyond the Milky Way, through the vast expense between galaxies, against all the odds, we've made it to intergalactic space. Out here, there's no horizon in sight. Even the closest galaxies are hundreds of thousands of light year away. The remains of galaxies are ripped apart by the Milky Way's huge gravitational pull, scattered among nothing or nothingness. This is as close as the universe gets to a perfect vacuum. But even this isn't totally empty. There are thin wisps of gas, fine traces of dust and something else. Is it "Dark Matter"? As per modern astronomy, it is so mysterious that we can't see it, feel it, taste it, touch it or even measure it. Yet so common, it could make up over 90% of all the matter in the universe. If dark matter does exist, it means there is no such thing as empty space (Make a note of that). Even out here, we're surrounded by matter. We think it exists because of its apparent hold on galaxies, like ours. Now,

This is the true edge of our home galaxy "Milky Way", a 6 billion year journey in today's fastest spacecraft, 160 thousands light years away from home. This galaxy should spin off into space but something is holding it here, something invisible, powerful, perhaps dark matter. Stars, clusters of stars, nebulae, it's a vast astronomical treasure trove. Before we left home, the universe seemed separate, something out there up in the sky. But now we know better. We are the universe, and it is within us. It's comforting to remember as we ventured through this abyss. Further and further, faster and faster. Now,

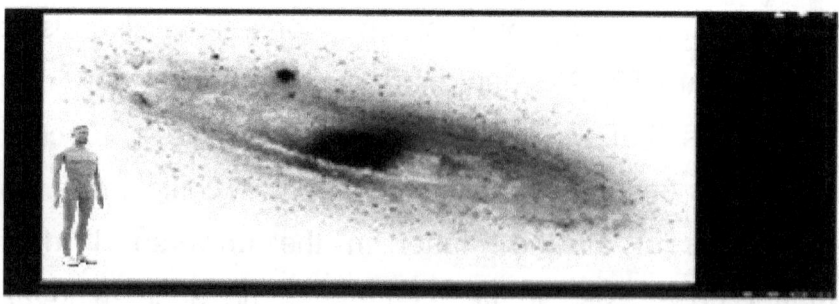

This is our Andromeda galaxy, a sister of Milky

Way, 2.5 million light years away. It's racing through space. Everything in this universe is blown apart, like shrapnel in an explosion because our universe is expanding with an ever accelerating pace; hence every element of it is going away from its neighbors. We are seeing this galaxy as it was when our ape like ancestors first walked on the African plains. Further through space and further back in time. There is an order in this chaos, a perfect pattern formation, behind the infinite variety, an endless cycle of birth and death, creation and destruction. It's a pattern woven through the vast fabric of space that binds each of these galaxies. Now,

$$10^{23} \text{ m} = 1,000 \text{ million light years.}$$

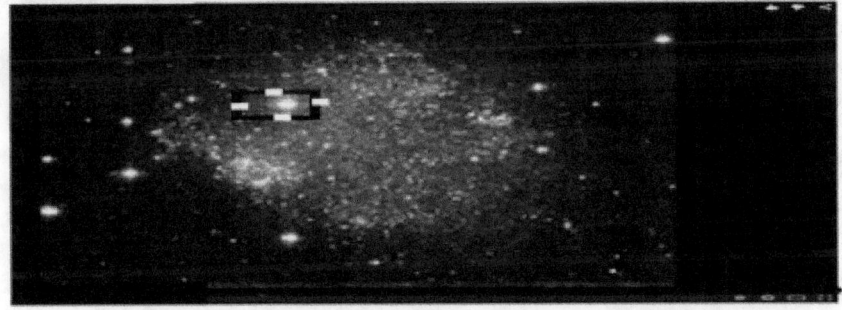

From this immemorial distance, all the galaxies look small, into a bunch of a family of galaxies including our Milky Way in a group, known as "Local Group". Ask somebody, where are you in this vast universe? And the next cosmic station for us is this, Local Super Cluster or "Virgo". Many clusters of galaxies like our Local Group are present in this single galactic supercluster. And it might look like this.

And next station for us would be the limits of observable universe. Now,

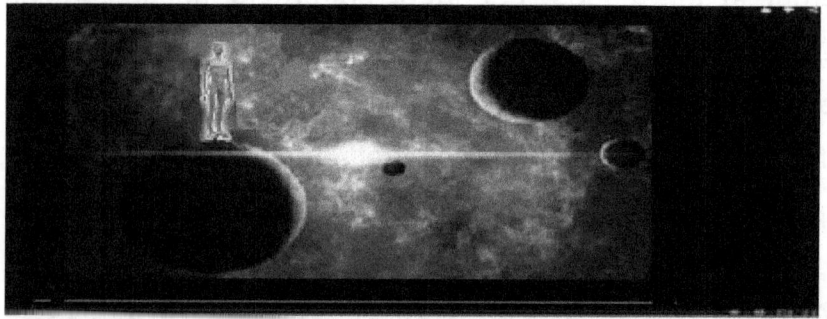

We are 2 billion light years from home, closing in on the edge of the universe and going back to the beginning of time, beyond Virgo as well as within it. This isn't a galaxy. Its brighter than the hundreds of galaxies. It's a blinding beam of energy surging for trillions of kilometers. It's a "Quasar" (short of Quasi Stellar object. Quasi? Yes it is quasi.), the deadliest thing in the universe. Our journey could be over, on this most powerful thing in the universe, a quasar. A swirling cauldron of superheated gas, this beast has a heart of darkness, a supermassive black hole, as heavy as billion suns. It's sucking the stars into nothing, lost forever from the visible universe. We need to go further, go faster. 8 billion light years from home, there are more galaxies but they look different, they are still young, still growing. We are getting close to where and how it all began. Look at the galaxies now; they look like primitive plankton in a dark ocean. It's like the clouds of dust and gas, merging to form embryonic galaxies. We have gone back, before the stars were born, into a cosmic dark edge and before the birth of light, the afterglow from the massive explosion, known as Big Bang that created the known universe. This is it. We have made it. This is the edge of the universe. 45.7 billion trillion km from home, 13.5 billion years ago, the very instant of the Big Bang. This is where our journey ends and where the universe begins. An infinitely hot, dense point erupts, creating space, time, matter and our universe itself. We are back, where it all started and only now we can enjoy our moment of creation and now this.

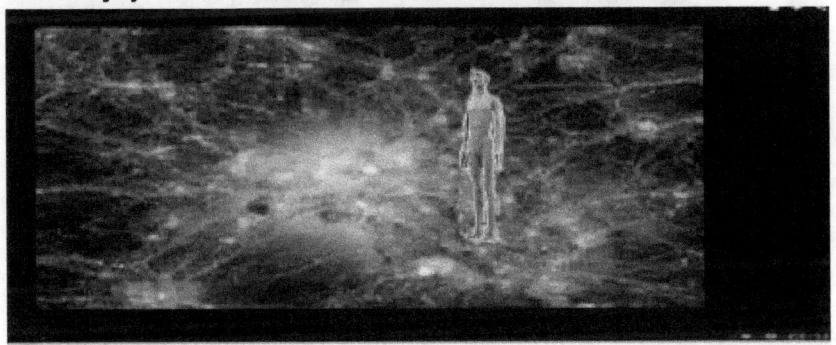

Even if you like to break the limits and go

beyond this observable universe, you would need to enter into this dark universe that lies beyond the reach of most advanced technology of today and might look like this.

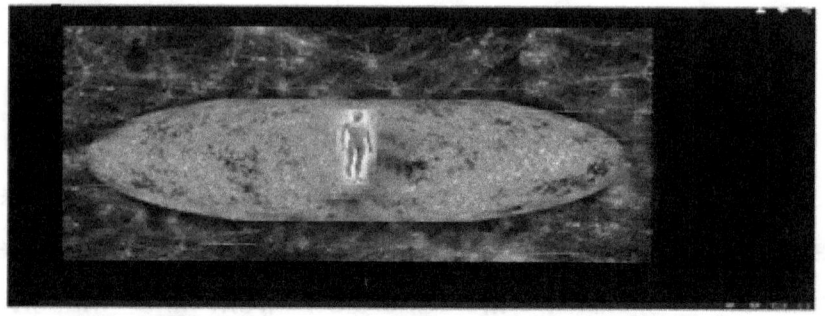

Congratulations. Finally, we have made it. You have touched the "Edge of the Universe". Now, wait for a second, before you go back to home. Look at this, an artistic perception about your existence, on this horizon of the observable universe, might be like this.

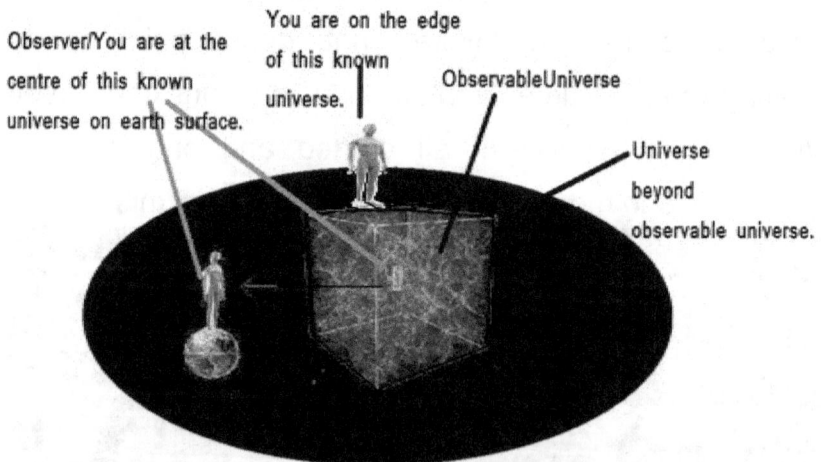

So, you can see that the question that was asked in the beginning is still unanswered because even when you are approximately 45.7 billion light years away from your home and even you are standing now on the edge of this universe. You feel that you are in the middle of nowhere. The question is as it was at the beginning that this darkness that is all around you when you are in your inky void is fundamentally the same as it is now. So, is this darkness really nothing? Because at this edge of the universe or observable universe, it is damn obvious that basically it is this darkness itself that is the one and only domain of this whole universe, it is the mother of this goddamn cosmos itself and it is the absolute domain of our absolute domain. So, it can't be "Nothing", but it has to be something and as a matter of fact it is but just wait for a while. Let the time come. [Well, it might be little tough for you but as a matter of fact; this is only half of our grand journey, in order to find the true edge of the universe.]. So, let's return back quickly. From 45 billion light years to $10^0 = 1$ m.

Plant   You are here

And this is it. Welcome back.

[Part 2]

And when we are back to the same point, where we started that is in our running metro. So, we will continue the remaining half of our ultimate journey but now in opposite direction. [And now we can have the tour plan for our remaining half of the journey too and here it is.]

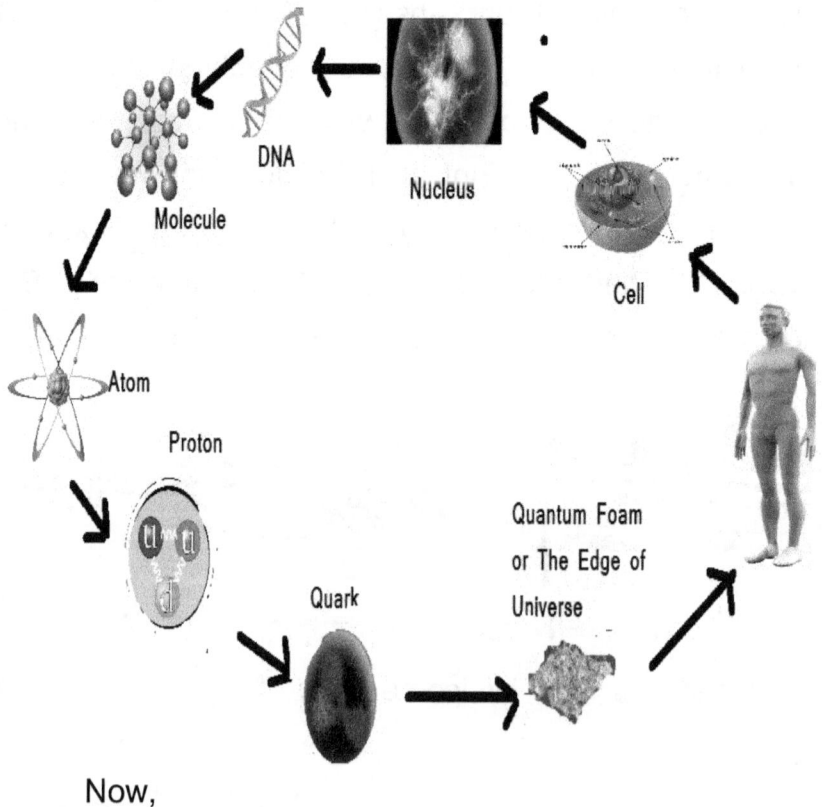

DNA

Nucleus

Molecule

Cell

Atom

Proton

Quantum Foam or The Edge of Universe

Quark

Now,

10^0 m =1 m.

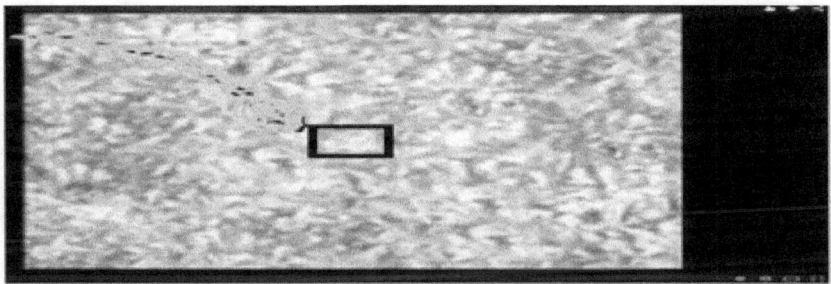

We are seeing the very surface of the same plant that is next to you on the floor of this metro train. We have started our trip forward but in reverse direction and we will go deep inside this small plant. Now,

10^{-1} m = 10 cm.

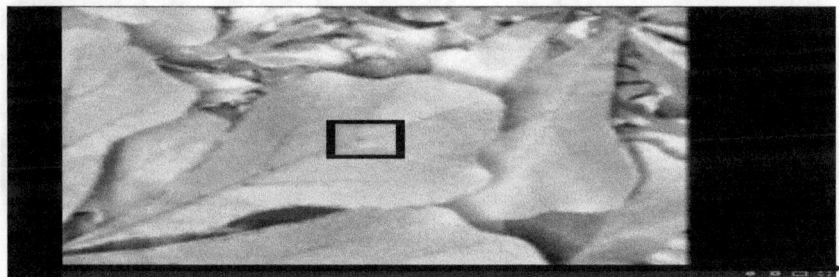

See, plant lines are clearly visible. So, our Journey towards infinitesimalness or what is called as "Quantum World" has begun. Now

10^{-2} m = 1 cm.

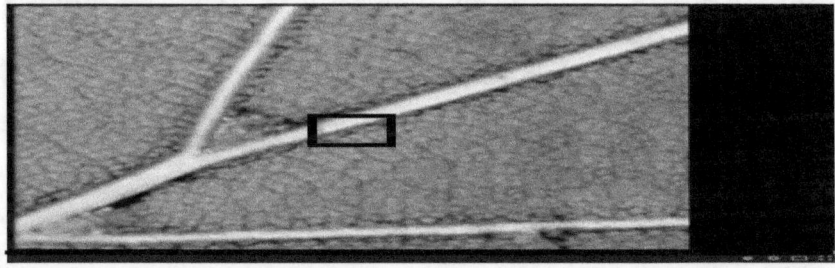

Divisions of this plant and its veins (at stomata level) are visible. Now,

10^{-3} m = 1 millimeter.

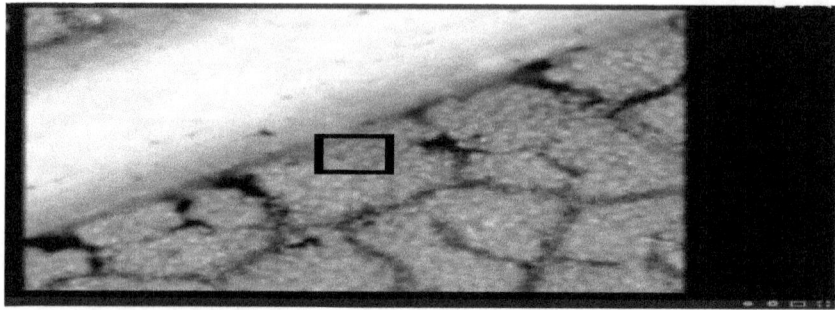

These are the cells of the plant. Now,

10^{-4} m = 0.0001 m.

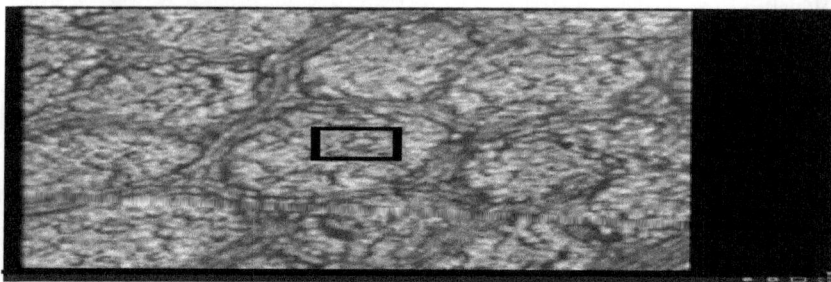

Cells seem clear and now,

10^{-5} m = 10 micrometer.

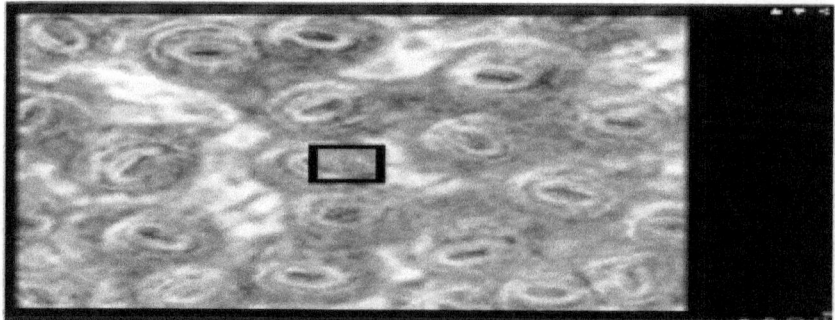

We have entered in the world of a cell. Now,

10^{-6} m = 1 micrometer.

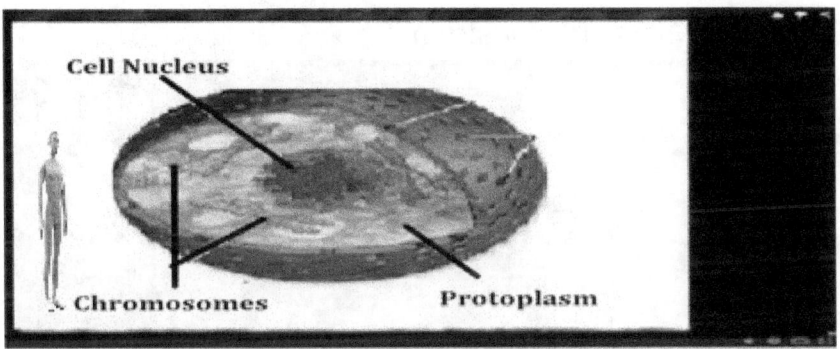

We are inside the protoplasm of the cell of the same plant. Now,

10^{-7} m =1000 angstrom.

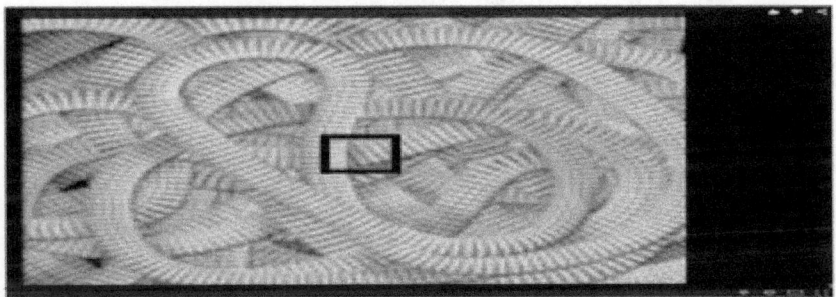

We are inside the nucleus of a cell of this plant and we are observing its chromosome. Now,

10^{-8} m = 100 angstrom.

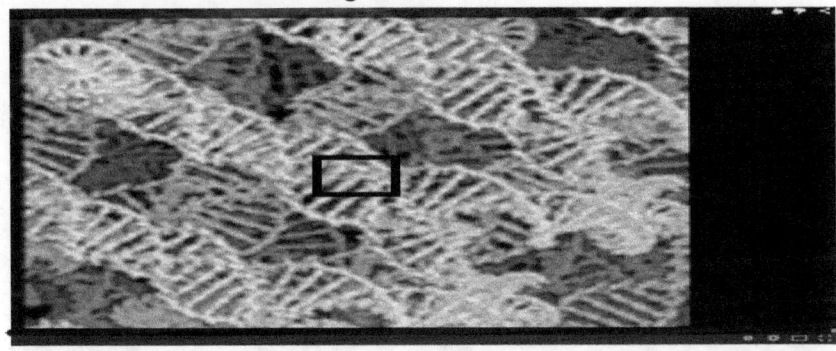

We are inside the chromosome and we are

looking at its DNA. Now,

10^{-9} m = 10 angstrom.

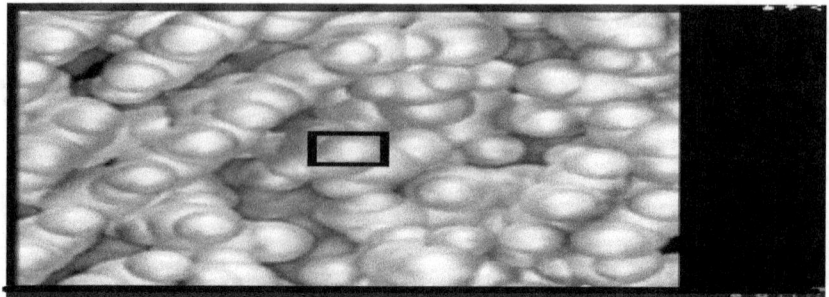

This is the molecular level of the same DNA. From here begins the study of genes. Now,

10^{-10} m = 1 angstrom.

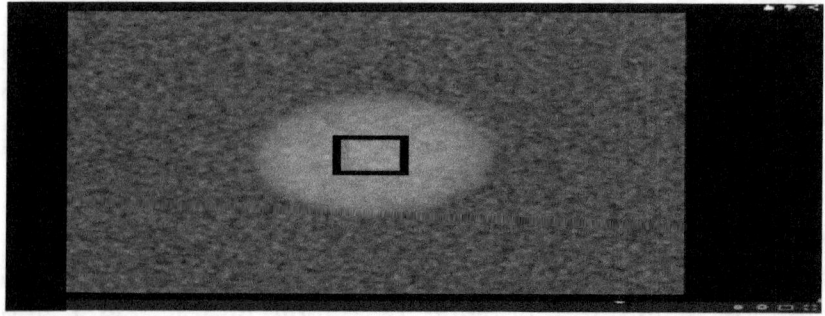

These are the atoms of the molecules of the same DNA, the fundamental unit of a living cell. And now,

10^{-11} m = 10 picometers.

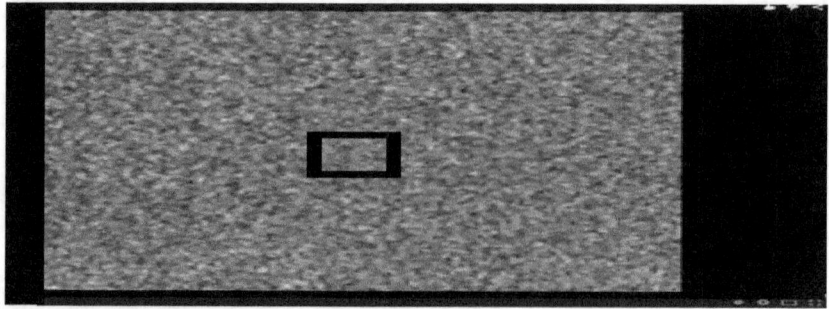

It might appear like we are in an electron cloud around the nucleus of an atom. Now,

10^{-12} m = 1 picometers.

As we saw, a large empty space within an atom. Now,

10^{-13} m = 0.1 picometers

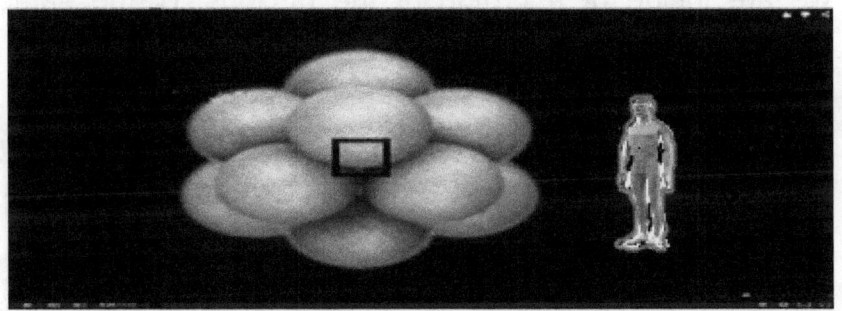

This is the centre of the nucleus of an atom of carbon, present in the DNA of the same plant. Now,

10^{-14} m = 0.00000000000001 m.

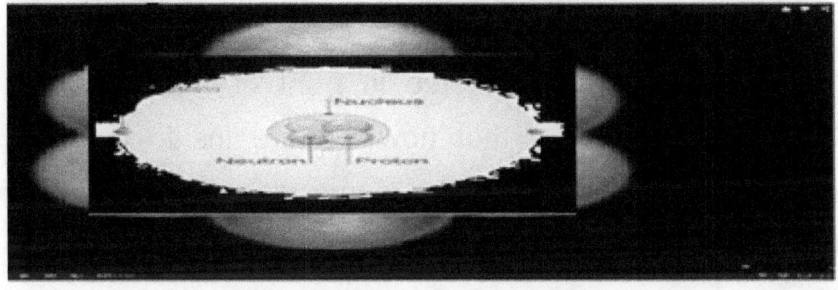

This is an imaginary picture of the centre of the

nucleus of the same carbon atom. Now,

$$10^{-15} \text{ m} = 0.000000000000001 \text{ m}.$$

We have entered in the science fiction experiments that demonstrated to be the face to face with a proton. Just to make it simple, we can say that we are inside of a proton of the nucleus, of the same carbon atom that is present in the DNA of the cells of the same plant. Now,

$$10^{-16} \text{ m} = 1 \text{ altometer}.$$

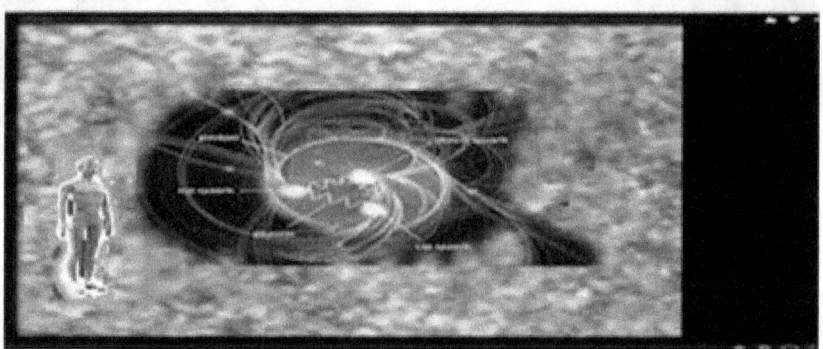

Here is where to step in front of what is called "Quark". Or, we say that now we are inside a quark, of the same proton, of the same carbon nucleus, of the same DNA, of our plant, on the floor of the metro train.

Do you remember, when we were 45.7 billion light years away from earth and we were standing at the edge of the observable universe, it was certain that we could go ahead and could continue our journey forward because there was no barrier out there. Well, if you can see, it is the same case here too. But if we go ahead our journey won't be scientific anymore. As it would become a subject matter of "Metaphysics".

So, again from this world of quarks, we need to go back to the same point from where we started this journey that is the floor of your metro train and here you are.

Plant   You are here

Well, as a matter of fact, the fundamental nature of our whole journey towards the cosmic realm and towards the quantum world can be summarized like this.

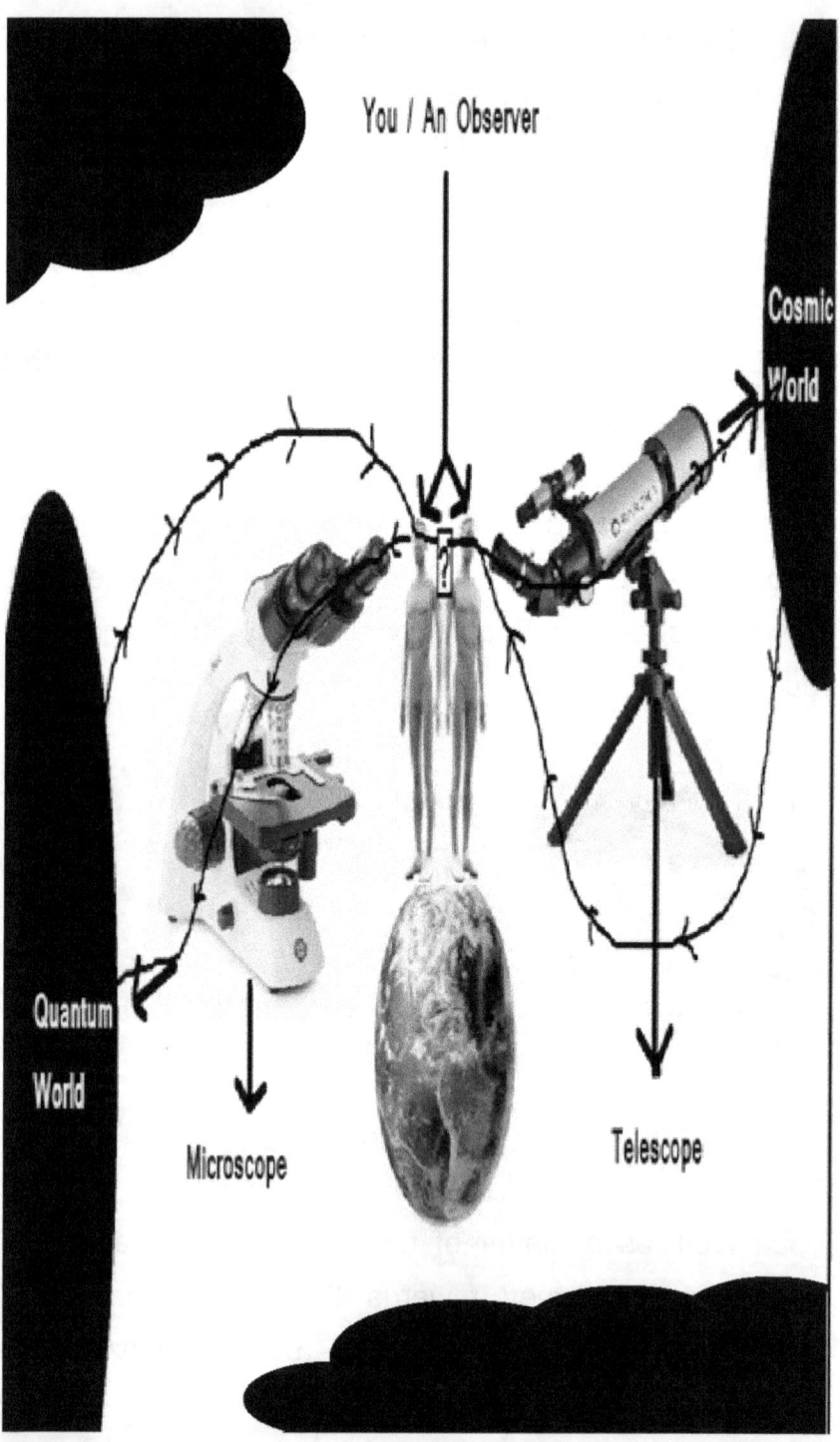

[Undoubtedly, you as well as I can see that it is quite probable that these points, first, when we are in a metro, second, when we are on the edge of the observable universe and third, when we are in the quantum world, they all are not different points but exactly similar to each other but there apparent isolation from each other is a resultant of the second order of absolute powers of the cosmos that is "Power of Surface". Does it imply that the whole of the observational illusion or platonic "Power of Appearance" passing through a telescope and through a microscope basically ends on same point?]

And here we complete our "Absolute Journey". So, what we've got? Nothing?

Well, I am so sorry; this all took a great deal of your time. I did what I did because there was a perfect reason for this and the reason was that I wanted to give you simplest and perfectly clear picture of what we are dealing with. Basically our true journey begins now.

[Part 3]

Your location must appear something like this, when you are boarding a metro train and travelling between two metro stations.

You are here

And it is exactly the same observation (but just without a metro train) that has been used since the beginning, in order to explain the idea of inertia and a jerk on your body.

Perhaps this might be an error as your actual position is on Earth surface, relative to which you do all the calculations, presuming it on at rest, what a physicist inside of you would like to call it as "Inertial Frame of Reference". Far from being true, your true position is this.

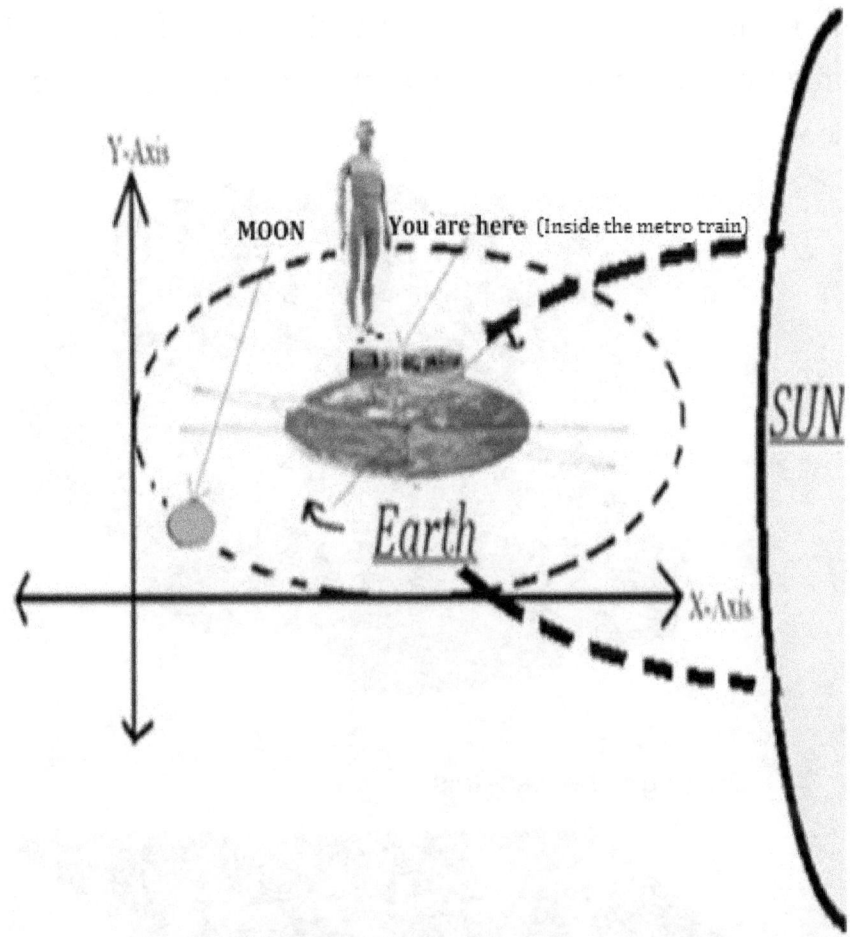

As you can see, without even realizing that the truth might be bigger than our Earth bound local observation, they all invented the science of inertia. But the problem doesn't end here as we know; we are in this dawn of 21st century human civilization. So, we do have a lot more information on this subject, for instance, now we go even farther and say that Earth is merely one planet among 8 other planets, starting from Mercury, Venus, Mars, Jupiter, Saturn, Uranus,

Neptune and Pluto, in the Solar System, so, the next true position of you as a true observer in nature, might look like this.

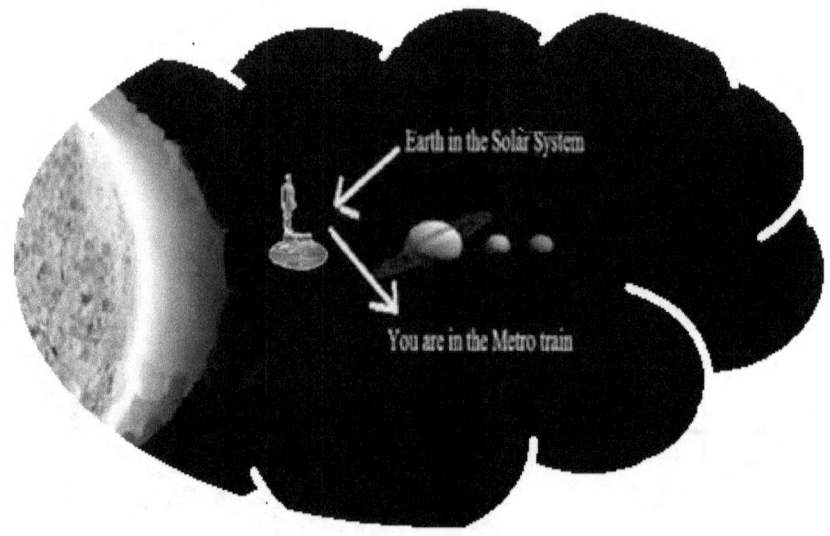

Or, might be like this.

Again, we are not done just yet, we need to continue our journey towards the true position of us as an observer in this creation or in the universe, in order to collect enough information, so that we could try to find out the truth of "Inertia", an inherent power of every single massive physical existence in nature.

So, now we go much farther, as we know that the Sun with its system is merely one among 200-400 billion (say 300 billion) stars present in our home galaxy "Milky Way" and we find that our true position inside our galaxy Milky Way might appear like this.

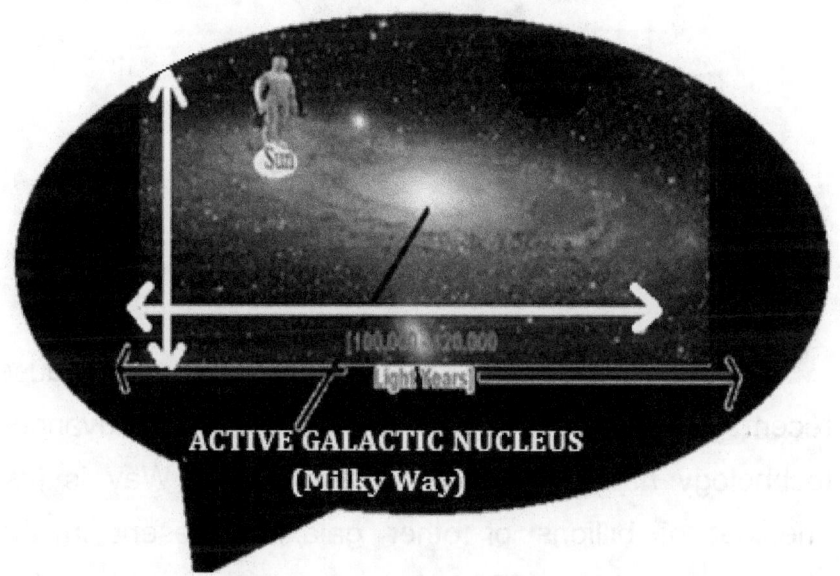

ACTIVE GALACTIC NUCLEUS
(Milky Way)

[Please note that there is a "Black Hole", exactly at the centre point of Active Galactic Nucleus (AGN) of our home galaxy Milky Way. Because there are lots of implications of this natural fact because modern observational astronomy says that approximately every individual galaxy in our universe contains its own Black Hole at its Active Galactic Nucleus (AGN).]

Or, your true position in this Celestial Coordinate System might look like this.

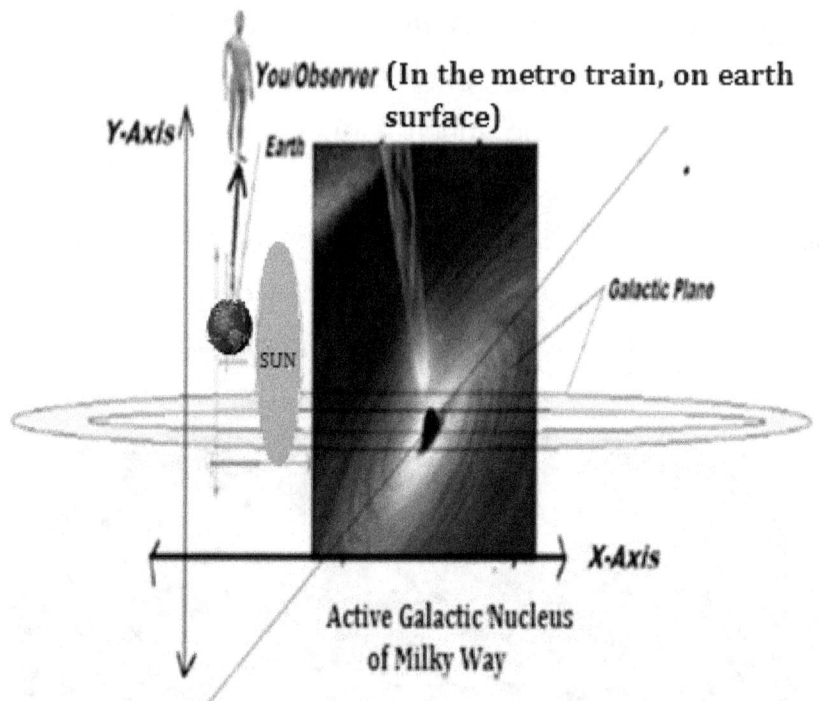

Still, we need to continue our journey because recent galactic surveys with the use of most advanced technology has confirmed that even Milky Way is just one out of billions of other galaxies present in the observable universe (Probably there are more than 170 billion galaxies in the observable universe and counting is still going on and this is the reason I would like to say that they all were out of time.) and Milky Way and her sisters like Andromeda (or M31) etc in fact are a part of a family that is called as "Local Group", which is a Galactic Cluster. Or, let's say, a new and what we might call true position of an observer within you (or somewhere else) in our own galactic cluster "Local Group", might look like this.

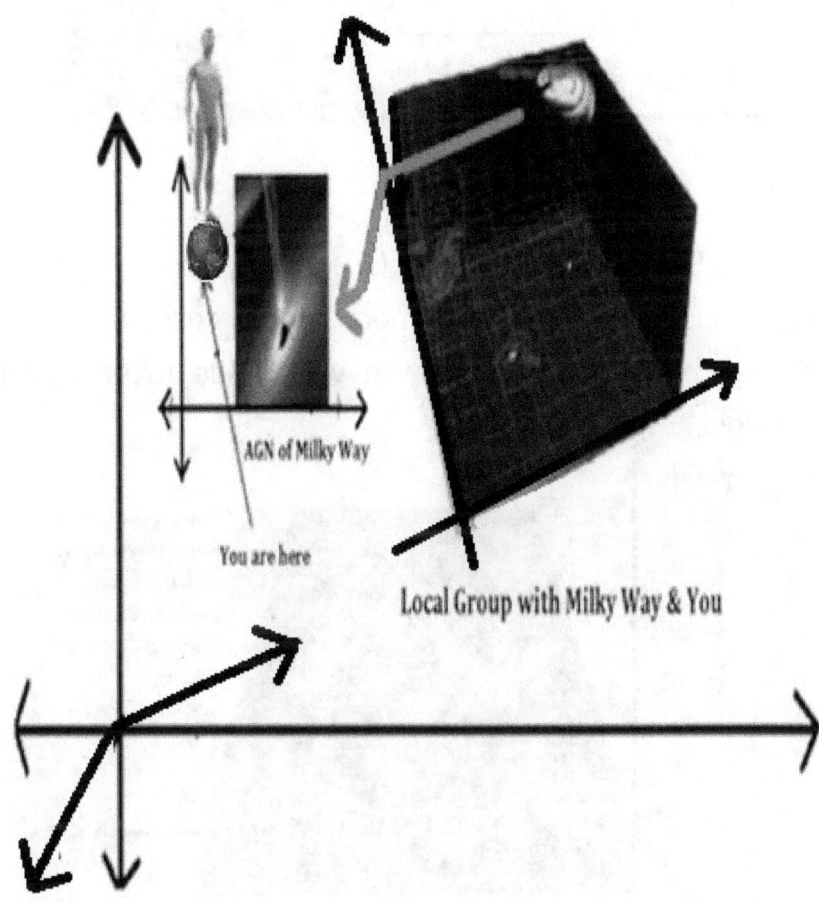

AGN of Milky Way

You are here

Local Group with Milky Way & You

But, Local Group (with Milky Way and you inside) itself is a part of a Galactic Super Cluster; we call "Local Super Cluster" or simply "Virgo". So, if we continue now, the true position of yours in Virgo Super Cluster of galaxies would be something like this.

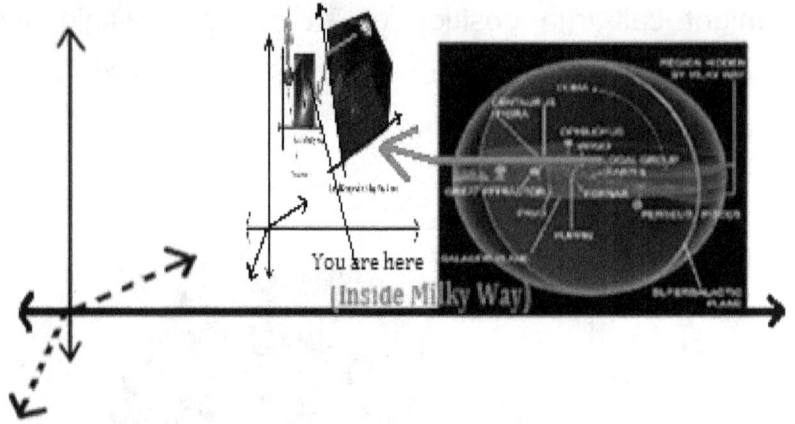

But Virgo itself is merely one galactic supercluster among many more in this cosmos. So, the existence of "Virgo" itself with you inside might appear like this.

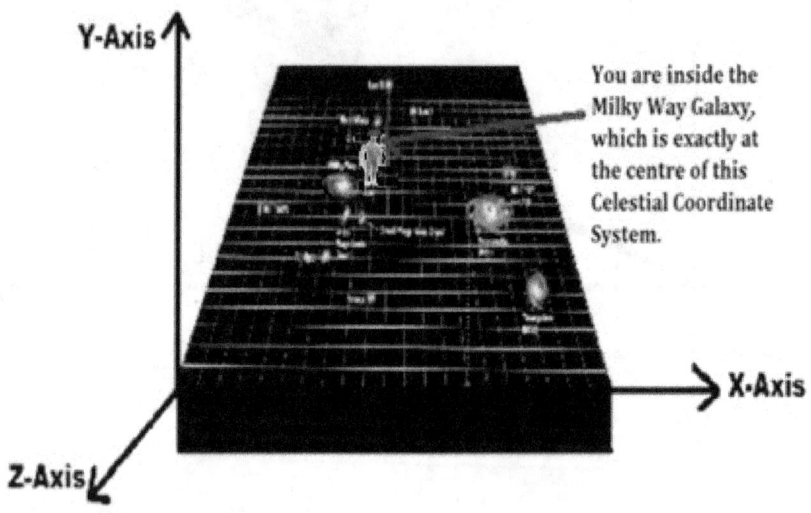

And today when with the advancement of technology, we know that the cosmic web of observable universe is made up of billions of galactic systems. So, the present scenario is something like this.

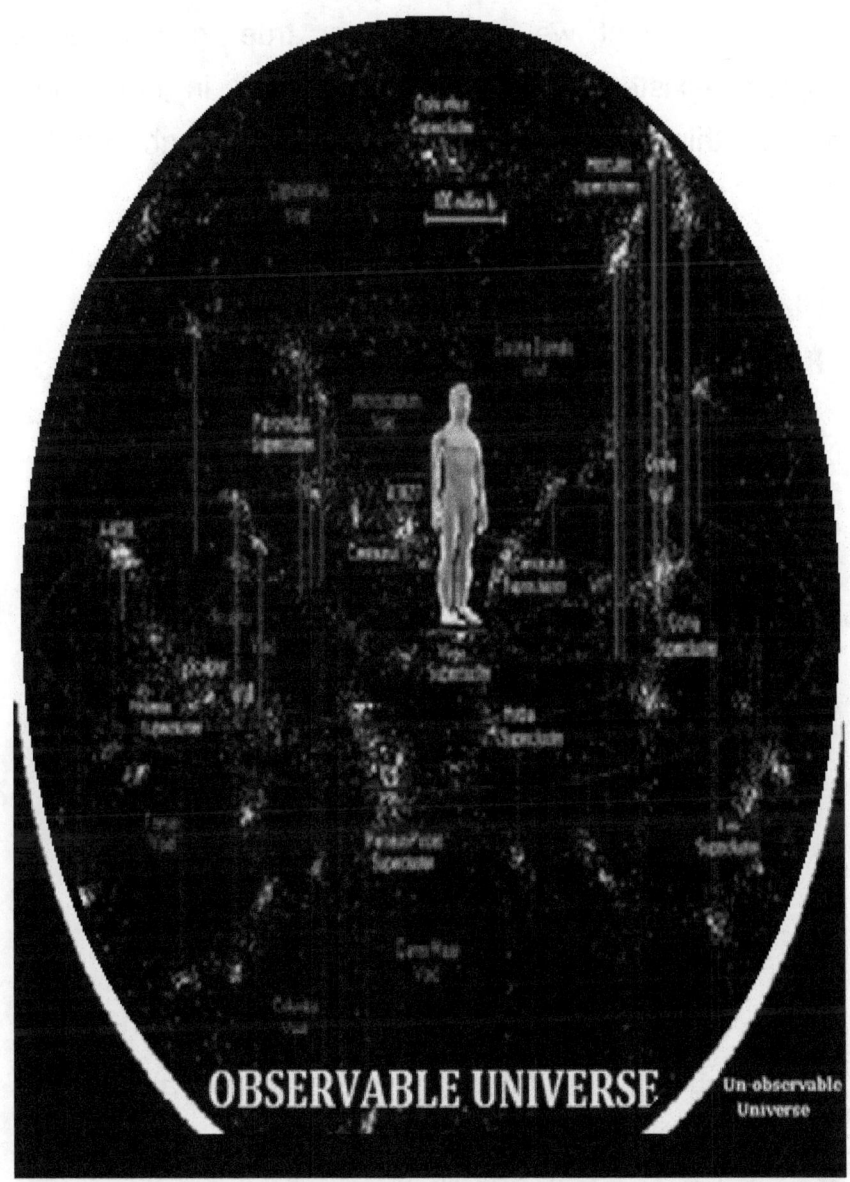

OBSERVABLE UNIVERSE   Un-observable Universe

[http://www.atlasoftheuniverse.com/nearsc.html]

[Just place yourself exactly at the centre point of this Celestial Coordinate System, next to Virgo and then you will realize what I am trying to show you since the beginning of this chapter.]

Or, what we call as your true location as a physical existence and as an observer in the nature and in this apparently infinite universe, might be like this.

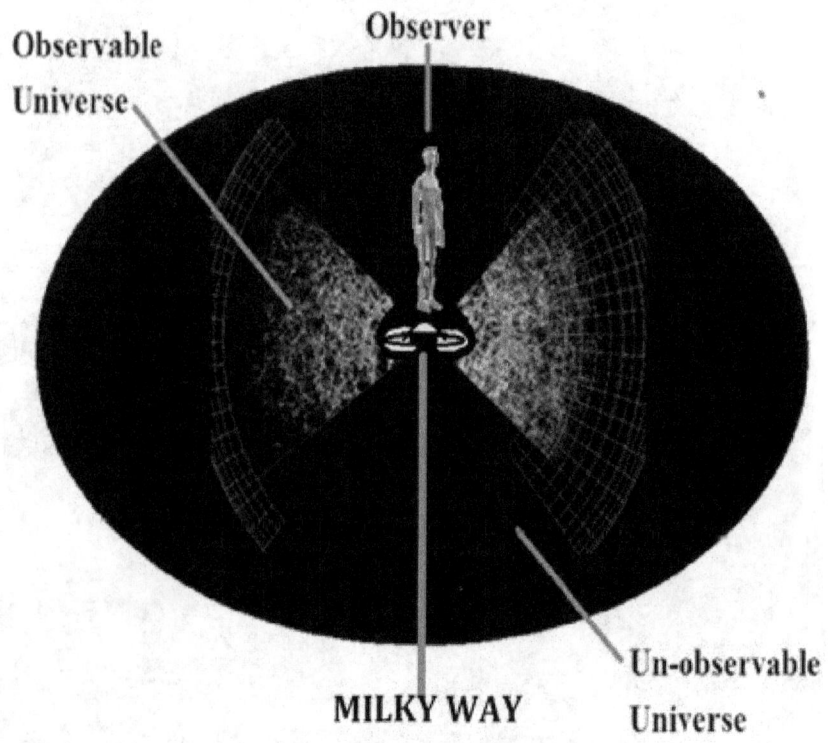

Observable Universe

Observer

Un-observable Universe

MILKY WAY

[You on the floor, in your metro train that is running on the surface of Earth (or even on the floor of your inky void, if you like) are exactly at the centre of this artistic perception of the observable universe.]

Even this might not be an end because there is no boundary out there that stops you, as you can see.

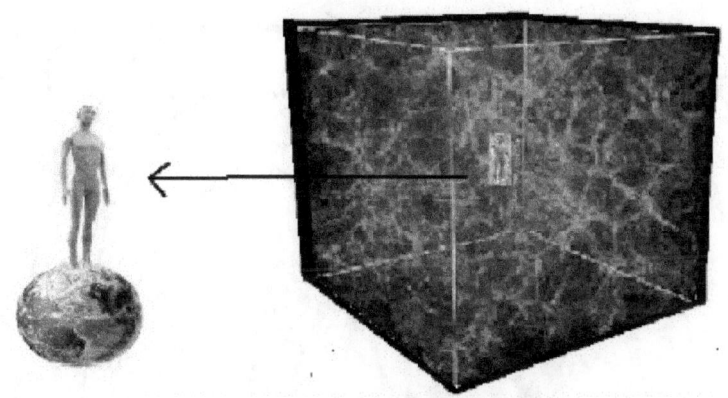

OBSERVABLE UNIVERSE
[You are exactly at the centre point of
this celestial coordinate system and
every single light point around you is a
galaxy.]

But we just can't go beyond, as of for now (perhaps just because of the same fact that it won't be scientific anymore).

So, the whole idea to show you this all was to give you a clear picture of what we call "true position/existence" of you as an observer in the universe or even more precisely, the true location of a physical entity or existence in this creation (of this universe or in this universe, depending on the faith you have either in Big Bang theory or Steady State theory). Similarly, if you would like to reinvestigate this super journey of yours in this universe than it might look like this.

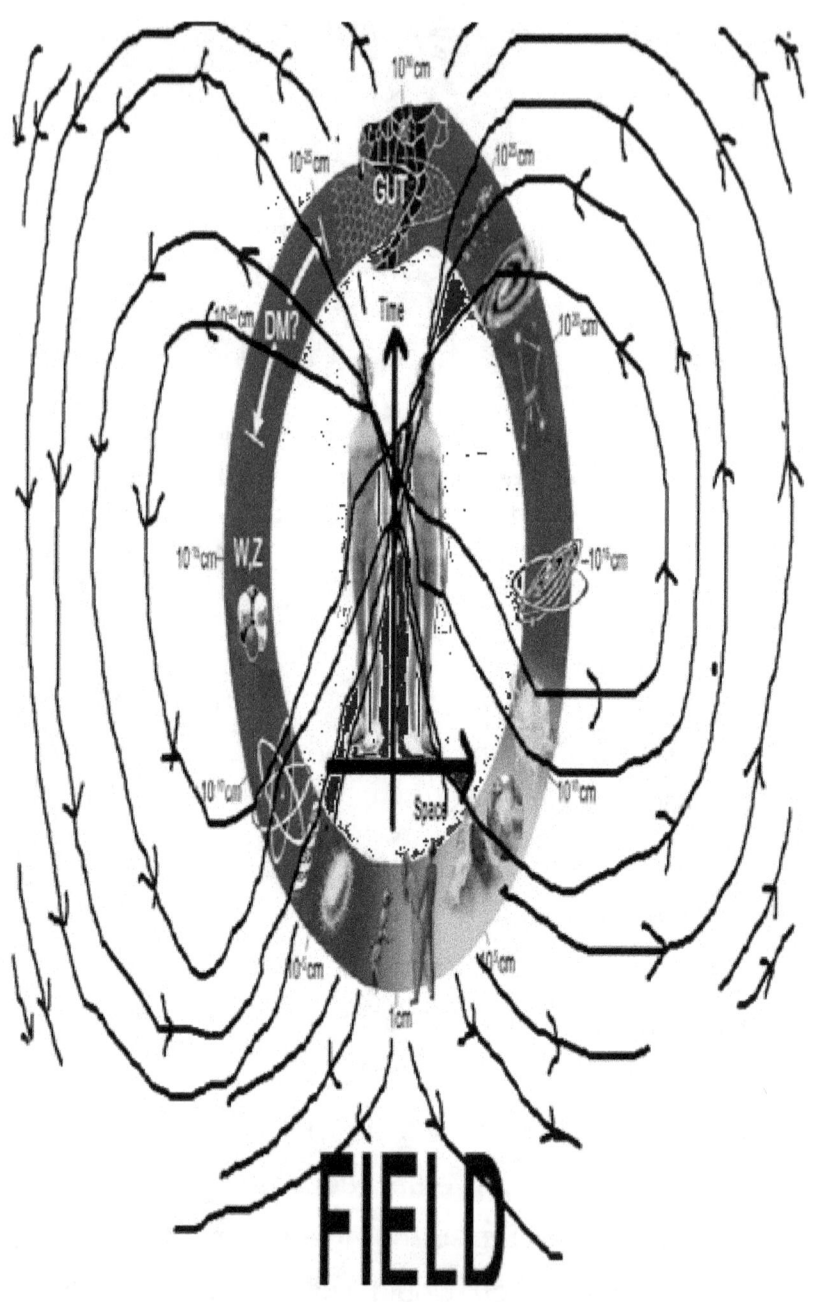

[D]

We need to go back, where we begun that is the floor of the metro train (or even the floor of the inky void, if you want to), where you felt a jerk. But now you may realize that it is a bit different, just because of the fact that now you are thinking universally.

Different in the sense that now you can see that your true existence as an observer in nature is not just a local, earth bound aspect as they thought but it has a perfect cosmic manifestation, which means that whatever you observe or experience here on earth, must have an intrinsic relationship with this natural fact but you don't feel it. Well, why not? Perhaps either because you don't want to or you think that it is too complicated for you and I am really sorry to say that. What Newtonians did was that they focused on the single observation, they were allowed to observe and then they tried to explain the science of the phenomenon, what we now call "Inertia" and as we saw, it was (and it is) far from being true, as you can see here.

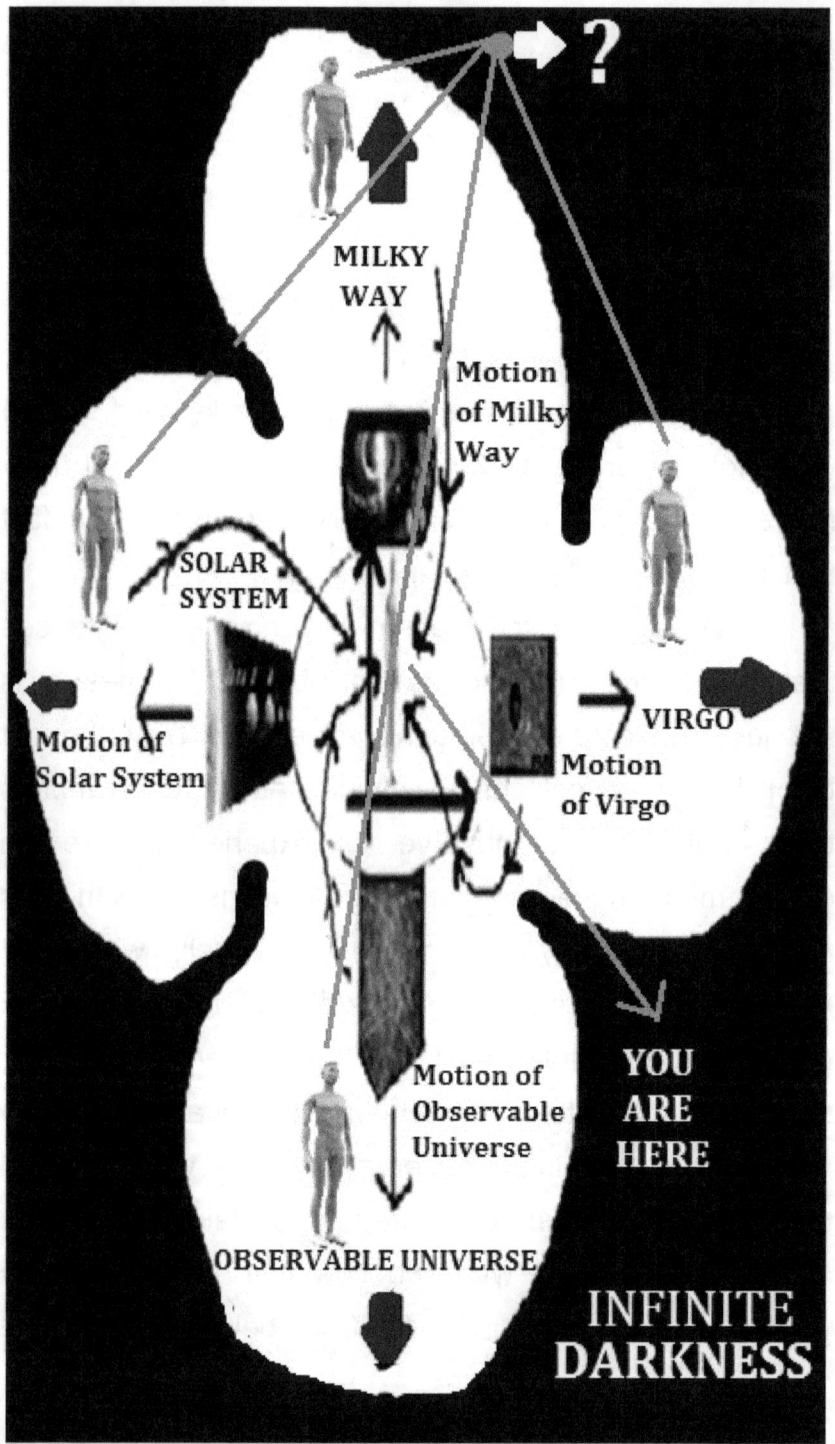

Now, the whole point is that when they decided to explain a natural phenomenon scientifically, they never had enough information about the system they were dealing with. But they thought that whatever axioms they had must work universally and as a consequence the explanation they gave us became science of today. [Make a note of that because we can find out that origin of entire Newtonian Mechanics is axiomatic. Axioms borrowed from Euclid (300 BC, Egypt) and his book "Elements". Well, you might ask, "What is an Axiom?" An axiom is a premise or a point of beginning of reasoning on any specific subject. Or,

"An axiom is a premise, as evident as to be accepted as true without controversy."]

But now, I am pretty sure that you do realize that in terms of a simple transformation of coordinate systems or more precisely frames of reference (or you can say even 'worlds', as we are taking whole creation in our consideration), they all had no idea, how far they were from the reality of the system and our point is that a misinformed reasoning about anything can't end up on a true scientific theory. So, a final conclusion says that there is something wrong within the existing scientific frame work of classical concept of inertia. Well, it is not just my realization only as there were others, in the history of science that had a feeling that there must be something that is missing from the science of inertia. Mach (1901) even proposed that this local phenomenon has a cosmic link, as a phrase of our A1 about the idea of "Mach Principle" reads.

"Mass out there, influences inertia here."

[Well, I tell you what, with all due respect, it is for sure that he never had any idea that what this "Out there" really implies. Because today we know that this "Out there" simply means, more than 170 billion other galaxies in outer space.] Years later, a well known scientists and a true genius also realized that it (Inertia) is something about which "Nobody knows that what it is" and this genius was Sir Richard Feynman, which he confessed in his TV interview.

So, the point is that, even if they say that everything is fine, we do have a perfect understanding about the truth of inertia and everything is done. All right but there is another well established and accepted theory known as "Probability Theory" and on the basis of this theory, let's say that there is mere a probability that something that has an intrinsic value is still missing and the "Mystery of Inertia" is not yet known and even if we fail at the end, it will at least bring more confidence in our existing understanding of what inertia really is. So, there is perfectly no harm, if we reinvestigate this well established principle of inertia, a bit deeper or does it? So, here we go.

When I was transforming coordinate systems from the floor of your metro train to earth surface, to solar system, to Milky Way, to Local group, to Local Supercluster or Virgo, to Observable Universe, [Even we could try to go beyond the observable universe but we didn't for a reason. Because modern astronomy says that some parts of the outer universe are too far away for the light emitted since the Big Bang to have had enough time to pass through the outer space and reach us in our telescopes on earth, so, these portions of our universe lie outside the limits of observable universe as of for now but in the future, (not too distant) light from distant galaxies will have had more time to travel through space and hence these distant parts of the universe will be observable on earth.]

[Please note that this fact might help us to realize that there is another probability that says that there is something wrong within our modern "Big Bang Cosmology".]

One thing that was continuously missing from our view was 'Motion', as "State of motion" of the object or its coordinate system or frame of reference as a whole is really important and as per the fact, not even a single physical object or even a coordinate

79

system or our reference frame in our whole consideration was (and is) on at rest (Far from being true what Ari thought.) like as if,

"Universe had already told to it's every single object that as long as you are inside of me, there is no way you can rest and you have to move all the time."

So, starting from the metro train and you inside, up to the limits of observable universe [let's say, it is approximately 90 billion light years across, as if you presume observable universe as a cosmic sphere, keeping yourself exactly at its centre, its estimated radius would be approximately 45.7 billion light years, as per the recent cosmic surveys measured by the most advanced technology like HST (Hubble Space Telescope)].

Every single co-ordinate system or our reference frame to observe an entity or an event in the cosmos is in motion.

They all are dynamic, as if there was or there is a mysterious law of nature under operation, just for this purpose.

Now, leave inertia for a while and see this new thing "Motion" that is everywhere, like a cosmic truth. So, beyond relativity, there must be a kind of motion within every single physical object of a particular reference frame, let's say it is

"ABSOLUTE MOTION".

It is present everywhere in a particular system and every object of that system shows this absolute motion that is beyond any sort of relativity. I think that it is now, when we must make our proto-hypothesis, so that, we will finalize our findings, at the end of this book. Agreed? Okay. So, if it is the case that there is an "Absolute Motion" and if there is an interaction in terms of a force on a physical object like your human body (either in your inky void or on the floor of the metro train), it must try to deviate it from the main course in which it is already moving (Towards its predefined destination? Well, Of course, this is the strongest probability). And if it is true,

It will show exactly the same kind of behavior as if there was or there is an inherent tendency or an intrinsic power, inside of it that resist any change in its state of motion. What we call "Inertia" (The same "Power of resistance", within an otherwise inert stuff).

So, the reason for this natural phenomenon when a physical object shows a resistance to any change in its state of motion (or even state of its existence), when a force is applied on it, still have deeper consequences. Even it appears that

A local perception of inertial nature of a physical existence in the cosmos might be a consequence of an intrinsic dynamism of the system to which we call "Creation".

Having said that everything in the cosmos is in a continuous motion (a sort of continuum of "Cosmic Dynamism") what we just called as 'Absolute Motion' and 'At Rest' itself is just a natural impossibility, we also need to see what we mean by this "Everything". And for this we need Quantum Mechanics, as every physical object like Earth or stars or galaxies or even your human body is made up of small particles universally. So, it will be no mistake, if we say that

Cosmic continuity that abhors "at rest", imposes a law of nature to every single physical or every individual existence, so that every single particle maintains a continuum of motion and it is this motion that in a certain reference frame doesn't have its relative counterpart and it is this reason that makes it absolute (But its nature is different from what Newtonian and Einstenian notions suggests).

It is true for everyone and perhaps the difference between these absolute and relative motions is that where relative motion in the universe is as obvious as the physical object itself but absolute motion is not an observable theme of the universe, quite similar to the situation that a person sitting next to you in a running train presumes that you are on at rest (relative to him). So, the conclusion is that

"Every single physical existence or a single physical entity or even every particle, no matter how big or small it is, no matter where it is in the universe, no matter how much mass it possesses inside of it, always remains in motion to which we called it as "Absolute Motion" and at rest or more precisely "Einstenian True At Rest" is mere a natural impossibility or a cosmic illusion.

That in turn as the history of evolution of modern science shows us today is far from being compatible with any Aristotelian or Galilean or Newtonian or even Einstenian institution. As bold as we can, we would say, they all were suffering from "IDS" that is "Information Deficiency Syndrome". Hence as a simple consequence, they all were misinformed as far as this particular subject is concerned and I got logic as well that says,

"Whatever you observe in nature or creation, you observe it, because and only because you are allowed to observe it." [As if, human senses like touch or vision to which everyone considers as gifts of God to humanity are themselves nothing but the "Absolute Limits".]

Consequently, a simple translational motion is perfectly a subject matter of human senses and hence it is observable but an intrinsic property of a physical or even every single particle of this universe is not simple to perceive.

And if you really want to enjoy the subject that what is observable and what is not. Or, what is a subject matter of your human senses and what is beyond it. You can use a brilliant idea of Plato, when he told us about "Power of Appearance".

"Power of appearance, led us astray and through us in confusion, whereas, art of measurement would have caused the soul to live in peace and quiet abiding in the truth."

Because it is really quite risky to challenge the existing establishment, what we call science of inertia or first law of Newtonian Dynamics. So, it is better if we re-examine our logic itself. Yes? Okay. And here comes a new thing.

[E]

I n 1970s, something strange happened in the field of science. Astronomers after completing their calculations came up with a bizarre idea. They said that there is one point in the universe, some 150 to 250 million light years away from us, in the direction of a galaxy cluster called as "Shapely". This point what they called "A Gravitational Anomaly" contains several thousands of masses of our home galaxy Milky Way and because universe is expanding, so, everything is in a flow that is called as "Hubble Flow".

But due to this point, what they termed as "The Great Attractor", we are dragged towards it and as a consequence because this drag causes a change in our motion, then our expected motion in Hubble flow, we are deviating from this Hubble flow and we have a different and our own "Peculiar Velocity" in the universe and the scenario might appear like this.

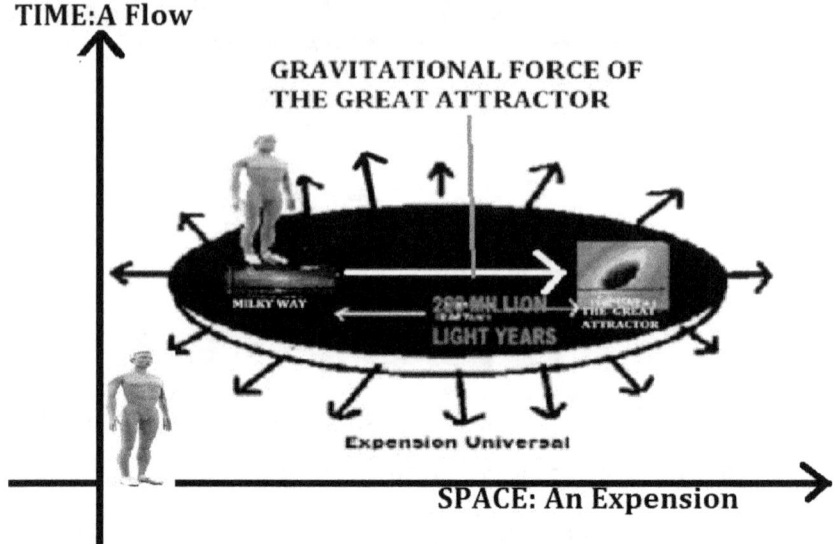

TIME:A Flow

GRAVITATIONAL FORCE OF
THE GREAT ATTRACTOR

SPACE: An Expension

And because like everything else around us, we are an element of this set, we call our galaxy Milky Way, so, we must also possess this 'Peculiar Velocity' here on earth surface, a sort of continuum of motion that is quite similar to what we just said is an "Absolute Motion" of every physical existence in the universe. That means every single observer like you or me (or even every single existence) in the universe always remains in a continuum of motion and this continuum never breaks up (something like an absolute stuff). The idea of great attractor has enough potential because it might give you a simple picture to imagine, what this absolute motion could look like and even Hubble flow and your very existence in the expansion of the universe do the same job. So, there won't be any harm if we presume that there is an absolute motion, due to which everything is in motion. But since it is like an intrinsic property of a physical (perhaps much more fundamental than any other), so, it is not just observable or what you say, not a subject matter of human senses.

Now, when a physical force or a classical force is acted upon a body, it basically attempts to change the main course of motion of a physical in the universe and as a consequence, inertia like behavior is observed, in a local frame of reference or in an earth bound coordinate system. But it is "An Illusion of Inertness" that simply doesn't exist. That in fact is the face of an "Absolute Cosmic Dynamism".

Due to its promising potential that it would give us some clues, let's talk about this "Great Attractor" a bit more. About 100 thousand galaxies including Milky Way are being sucked towards a region of space that we can't see and we don't know why. [Quite like a "Black Hole" that attracts everything that is in its vicinity and swallows it up.]

Astronomers call it a 'Gravity Anomaly' and as this "Great Attractor" is dragging us, so, your existence in this cosmic phenomenon might look like this.

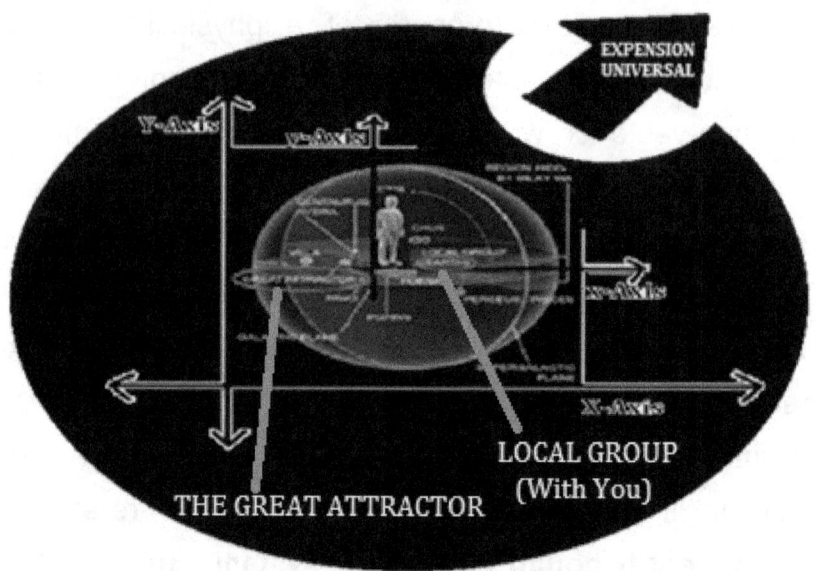

[Idea is simply to show you that say Milky Way with you inside, is near the centre of this diagram and the "The Great Attractor" is at some distance away, so, dragging everything including you towards itself.]

Here is one way to realize, how great this attractor is. As we said, there are 200 to 400 (say 300) billion stars in Milky Way, astronomers believe that whatever is dragging us has a mass of a million billion suns and it is approximately 220 million light years away. We know it is there because we feel its attraction force here. The universe is expanding with an ever accelerating pace that means all the galaxies in the universe should be getting farther away from each other except few exceptions like M-31 (that appears like a flow of empty space). This apparent movement of galaxies in relation to each other as the space between them grows that is called as 'Hubble Flow' is quite similar to any flow, like flow of water. But since mid 1970s, we know that we are moving with a specific rate in a specific direction, in association with this apparent movement. This deviation from Hubble flow is called object's 'Peculiar velocity' and in our case, the object is our Milky Way galaxy (Well, we could say that the object is you). By the 1970s, when it was clear that we can measure our peculiar velocity by tracking the movement of the Sun through Milky Way with reference to the universe's Cosmic Microwave Background (CMB) and it turned out that we are moving with a velocity of 600 km/sec, towards the constellation "Centaurs". Now the reason why we need to talk about this Great Attractor is that

"If it is the sole source of the peculiar velocity of Milky Way, as a cosmic particle, this motion must be somehow related to an observer that exists within it, here on earth. And it is this clue that could help us in order to visualize the idea of Absolute Motion and its connection with every single physical existence that is present here on earth, including you and me.

If the Earth is orbiting the sun with 600 km/s, a year would only be 18 days long. Now, what is actually closing the distance between us and the Great Attractor, it is because it caught up in the Hubble flow too and moving away from us but much more slowly than it should be. So, obviously we need to figure out what it is. At first astronomers thought it has to do with a fact that the Milky Way is on the out skirt of a broader neighborhood of galaxies known as 'Virgo' Super cluster', so, maybe we are drawn in by gravitation of Virgo but even though Virgo has its galaxies, it is not nearly massive enough to drag in that fast. This means that

"There must be something even bigger behind it."

As a result of continuous efforts, by the early 1980s, it was confirmed that it is not just us that are moving towards it but everyone in its surroundings is moving towards this thing, that is hundreds of millions light years away but we can't see what we are heading into because we can't see about 20% of the universe around us because the geometry of Milky Way, as we will see, is such that our own galaxy is blocking our view, what is call as "Zone of Avoidance" and it just happened that this enormous thing is in this

20% behind the plane of our galaxy, in this zone of avoidance and it is so chocked by mass and dust that we can only see anything behind it, by searching radiations like X-rays. But this doesn't give enough clear picture of the truth. By 1990s those X-ray surveys revealed to us a centre of mass 220 million light years away, near another super cluster but then in the mid 2000, they discovered something that is even more massive behind that. This object called "Shapely" Supercluster is approximately thrice as distant as this "Great Attractor" from us and it has an estimated mass of 10 thousand Milky Way galaxies, that is a bigger size then our Virgo super cluster and this is the reason, it is believed that it is the most massive thing in the entire observable universe.

So, what is giving us our peculiar velocity? Is it Shapely or the great attractor or something even more massive and farther away? We don't know, nobody knows. All we know is that we are drawn towards an area of this hyper massive space and that is the reason, it is called as 'The Great Attractor', as it is really great to comprehend for anyone on Earth. So, to summarize this whole concept, let's draw a picture of it, when this Great Attractor is exerting a gravitational force on you.

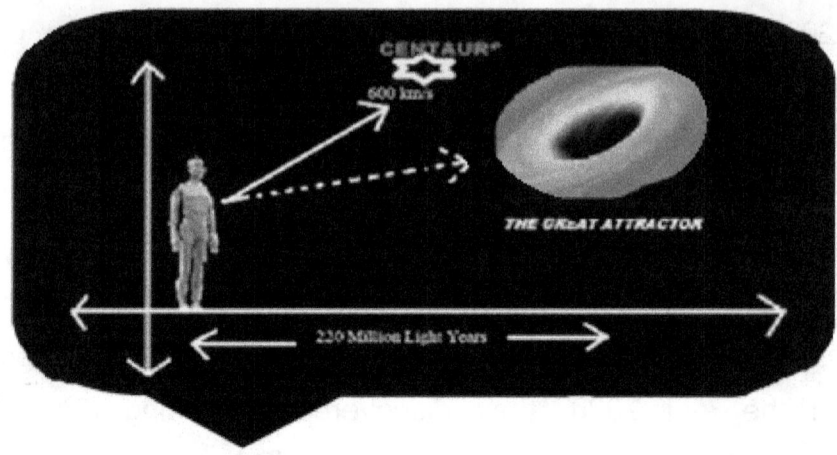

This whole idea is probably efficient enough to convince you that you (either in your inky void or on the floor of the metro train on earth) are in a flow, a cosmic flow to be precise, caused by either a drag of the Great Attractor or expansion of the universe itself. But you don't realize it at all, perhaps because everything around you is flowing, in the same flow with you and because of your limit or your observational bias that you always focus on the local aspect of every single observation or event.

So, now we need to deviate a bit from our main course because we need to talk about another scientific concept that might in turn will help us, to give some more clues to find out, what we are looking for (Well, what we are looking for? inertia. Great, you have it in your mind).

GREEN BOOK

[F]

Let me presume that you perfectly understand the intrinsic link between the concept of inertia and idea of mass as you do realize it all the time that it is not so easy to move the Earth to and fro but it is quite easier to set a tennis ball in motion. More mass simply implies to you to expect more inertial nature from that object in terms of its more resistance to change in its state of motion. Now, we need to investigate another thing that is one of the most advanced themes in this more than 3000 years long evolution of modern science and the human civilization itself. Let's imagine that you got a way out that is you found out, how to travel between your inky void to your real void or real world that is full of physical stuff, what you call your natural or physical world. So, when you are back, whatever happened with you in the inky void, you still have its memories and suddenly you pick a stone. Now, look at this, a stone is in your hand and you are looking at it as you never saw such a thing before in your whole life.

[That may be due to the fact that you are back from an inky void. Well, just to emphasize this point, we can see that the newness of the same stone is not real. But it is looking something new to us when we are back from a place, where we have felt its absence and it is this absence of everything in our inky void that is now creating sense to everything, when we are back in this physical world. Perhaps it is quite a similar situation that whenever we lose something or someone and when we feel its absence then only we truly appreciate its real importance.

For instance, suppose one day you go to a doctor and he informs you that you have a quite advanced stage of leukemia and you will live for not more than six months. And when you come out of the chamber and if somebody offers you a glass of water then you start wondering, what the hell it is, in this glass that goes inside of me. It is just water, as it always was but perhaps at this moment of your life, it is the biggest puzzle for you, why? Because now you are feeling the absence of "Life" and it is life itself that creates sense to everything, including water. As a matter of fact this is a really good point for us in order to find out the validity of "Einstenian Total Relativism" about that we will talk later on.]

So, when a stone is in your hand, you see that it also shows the same nature as your body did in the inky void that is if you want to put it in motion you need to use your hand muscles, to exert a force on it, to set it in motion. Why? Because it is also inertial, why? Because it possesses mass, the same stuff you have, in terms of your human body. All right, but

What the hell it is exactly? Because whatever it is, it is the source of your own physical experience and even the very existence of this physical world and the very source of solidity itself.

Now, before we go ahead, we need to realize that in ancient times and in approximately every religion, the idea of "Illusion of solidity", in other words the belief that "This physical world is an illusion", as Plato said in his "Allegory of cave" that we are nothing but shadows, used to be an intense area of discussion among philosophers, religious people and scientists and in time, science with its advancement, made it approximately clear that it is not the case and before this time, this point of discussion was almost dead and it is quite interesting that modern science as it evolved with all of its advancement and even the most expensive area of research of the day has finally come back to exactly on the same point. But now, what science is concluding is exactly opposite to what it used to say that is the results obtained from particle accelerators predict "There is No Mass" and no mass simply implies that there is nothing solid. So, were ancient people right? (Without wasting a great deal of time and money as modern science did) You see, when the stone is in your hand, you feel it is solid but if you ask to a particle physicist, he would say that it is made up of mostly empty space and when you apply a mechanical force on it, in order to press it even harder, you experience its solidity even more. Now, the obvious question is if this stone like any other physical stuff of your world is really made up of just empty space, where this solidity is coming from? Because it's this thing, that makes your world perfectly physical.

So, that is why you need to go back in your inky void and start thinking all over again, about your present domain because whatever is happening or happens, it is happening or happens in a void as its fundamental domain. So, it is this void that could give us some clues in order to resolve, what would otherwise appear as "Illusion of Solidity". And this is the reason we need to find out the truth of this void or emptiness or empty space.

Now, what we need to talk about is "The Mystery of Empty Space", what we perceive in the form of vacuum or void or nothingness or simply emptiness and the power inherent in this concept lies in the fact that it is ubiquitous, present everywhere in the universe, as the universe is made up of this emptiness only. Of course, philosophers even way back then Aristotelian time, have discussed a lot about it, simply because they realized its true importance but after that people became engaged in other research works and in 18th century when Mick (Michael Faraday, 1791-1867, England) came in, he reinvented the idea of the "Field" that is quite closer to this idea of void or empty space. After him came James (James Clerk Maxwell, 1831-1879, Scotland), who even gave us a mathematical viewpoint to look at it.

And now surprisingly modern science with all of its technological advancement today has a lot to talk about this emptiness, which is of huge importance because it is omnipresent in nature and what is more surprising is the fact that today it is one of the most intense area of investigation on the globe, with a realization that it could be a future source of energy. Even our A1 said so,

"Vacuum could be the source of energy."

So, the whole point is that how important this nothingness of nature or universe has become for entire humanity. Now, the question is "What this emptiness or empty space or nothingness or void or vacuum is? And in order to find out the answers, there are two entirely different divisions of modern science that are under operation today. On one side, we have world's most powerful telescopes like HST (Hubble Space Telescope) and observatories, where astronomers are investigating the flashes of light coming from far reaches of outer space, like distant Quasars and Supernova and they are trying to understand the nature of the vacuum in space out there and the second division of science known today as "Big Science" [Just because it is the most expensive area of research ongoing on Earth today like particle accelerators, as it is in CERN (European Organization for Nuclear Research), which is a billion of dollars project], where particle experimentalists are trying to create tiny ripples in space-time fabric, what is called as Higgs bosons (as it is the proposed particle of Higgs field that fills this emptiness or nothingness everywhere, as per the theory) and again the purpose is common, to understand what this emptiness is in this cosmos or in this creation. Let me tell you what they really are doing out there briefly, so that you might have a clear idea of this whole. First, the fundamental is that if you place a charged particle like a proton in an electromagnetic field, you can accelerate this particle. *[Make a note of that as if there is an interaction between field and a particle, it makes it moving faster because this interaction between a field and a particle is really important to us and we will need it in Blue Book.]* So, you place a charged particle in a particle accelerator and switch on the electromagnetic field and now this particle is moving with a great velocity and in the exactly opposite direction you place another charged particle and do the same thing, so, it is also moving with a great speed but in opposite direction and at the centre point they collide to each other and break down into smaller and smaller particles and it is these particles that tell us that what your original particle was made up of. And if you make your machine more and more powerful, by adding more powerful electromagnets and if you collide these two particles with more and more energy and can get to know what exactly they are made up of and it is this principle they are using there. It's simple. Yes? But what is impressive is their construction. LHC (Large Hadron Collider) as it is called is a huge machine at CERN that is built in a circular shape with a circumference of 27 Km and its tunnel is 3.8 meter in diameter. So, you can see, it is the largest machine that was ever made in the whole history of human civilization. First they tried to collide electrons and its antiparticle positrons and found some good results but as we are much more interested to discover 'Higgs Boson', so, they realized that they need higher energy particle collision and then they collided proton and anti-proton and this particle collision creates what is called as 'Fire Ball' (an exact replica of Big Bang, that is presumed to be the moment of creation of this universe but on a quite smaller scale) and situation might look something like this.

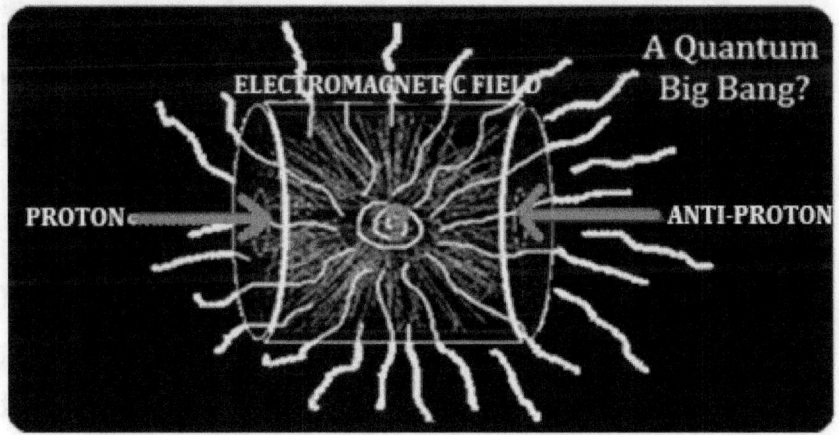

And in 2012, they claimed that they have found what they were looking for, "The God Particle".

[G]

As per the tradition, if you want to know what is the most fundamental building block of nature, what you do is that you take a piece of something and smash it, it gets braked down into smaller pieces and you realize that this was made up of these pieces of matter and again you smash one of these pieces, with more energy and this piece gets break down into even smaller pieces of matter and so on, till the end when you can't smash it more harder or you could not go beyond a certain limit and on this limit, you conclude that perhaps it is the most fundamental stuff of the nature. Same is the case with these particle accelerators, except the fact that because we have an advanced technology, so, we can smash a piece of matter with more and more energy that is far beyond our previous limits. Let's make it more simple, you take a block of ice and you smash it, it gets transformed into water and at this point you conclude that ice is made up of water and now you smash it again, even harder this time and let's say water gets braked down into water molecules and you say that ice is made up of molecules and you smash water molecules now and it gets break down into atoms of hydrogen and oxygen and at this point you conclude that ice as a piece of matter is made up of atoms and as the idea of "Atomism" from John (John Dalton 1800, England) shows, it used to be our limit. So, in the beginning of 19th century, we said that the most fundamental stuff of nature or universe is the atom. And the world is made up of atoms and even today some people believe that it is the truth. But when we realized the fact that a field like an Electromagnetic Field can accelerate a particle, what Thompson (J.J.Thompson 1856-1940, England) used in order to make his brilliant and Nobel prize (1906) winning discovery of "Electron", they made even more powerful smashers and when atom that was our previous limit was smashed, it got broken down into protons and neutrons and at one time we again thought that the world is made up of these protons and neutrons as the most fundamental particles. But when proton was smashed, it got broken down into "Quarks" and now it is our present limit. So, you can see that it is like as if you open one particle, you find there are other particles and an empty space inside of it and if you go ahead and open up one of these particles again, you find that there are other particles and same empty space inside of it ....ad infinitum?

Like what if you get a box as your birthday gift and you open it up and you find that there is one more box within it and because you are pretty curious to know what is your birthday gift, so, you go ahead and open this box now too and again you find that there is one more box within it and so on. What you think? This "Particle within the Particle" or "Box within the Box" conundrum will continue even beyond the quark level? As if we find in future that there is "Pre-quark" that makes up quark (Or even a "Super Quark" that makes up Pre-quark itself) or is it "Quark" itself that is the most fundamental stuff of the universe? Only the future will tell. Well, wait a minute, you don't think that it is strange that there is a particle within the particle, within the particle, within the particle, that is quite similar to a puzzle of a box within a box, with in a box, within a box...ad infinitum? Or, does it ends somewhere? Because if you recall, Ari told us that it will keep on doing it, without a limit.

This whole idea is good enough to show you that for the quest of ultimate knowledge about this creation that has been started since the dawn of human civilization, with the technological advancement, we have removed one or more layers from the ultimate truth, in order to dig up the final layer of knowledge of this universe. But still there are layers, yet to be uncovered and even today also we don't have any true idea that how far or closer we are from this absolute layer of the ultimate truth of the universe or in other words, even with the help of most powerful particle accelerators like LHC of CERN, we still have absolutely no idea that how many layers are there that are yet to be removed. So, is this all proves that Ari was right after all?

You know what, there is a text or I should say an ocean of text, whose origin is unknown and mysterious as well. But what is for sure is that it was written thousands of years ago and since then it is with us and what even more surprising is this.

न्यग्रोधफलमत आहरेति" "इदं भगवइति' 'भिन्धीति" भिन्नं
भगवइति" "किमत्रपश्यसीति" "अण्व्य इवेमा धाना भगवइति'
आसामङ्गैकां भिन्धीति" "भिन्ना भगवइति" "किमत्र पश्यसि
"किञ्चन न भगव इति' त ৬ होवाच ॥ १ ॥

इस (बड़े के वृक्ष) से बड़ का फल लाओ।

यह है हे भगवन्!

तोड़ दिया है हे भगवन्!

इसमें क्या देखते हो? ॥

बड़े सूक्ष्म से दाने है भगवन्!

प्यारे इन (दोनों) में से एक को तोड़ो॥

तोड़ दिया है हे भगवन्!

इसमें क्या देखते हो॥

कुछ नहीं, हे भगवन्!

'यं सोम्यैतमणिमानं न निभालयस एतस्यैव सोम्यैषोऽणिम एवं
महान्यग्रोधस्तिष्ठति ॥ २ ॥

उसको उसने कहा हे सोम्य! तू अब जिस सूक्ष्मता को नहीं देखता है, इसी
सूक्ष्मता से हे सोम्य! यह इतना बड़ा बड़ का वृक्ष खड़ा हो जाता है।

श्रद्धत्स्व सोम्येति स य एषोऽणिमैतदात्म्यमिद ৬ संर्वं तत्सत्यঽस
आत्मा तत्त्वमसि श्वेतकेतो! इति। भूयएव मा भगवान् विज्ञापयत्विति।
तथा सोम्येति होवाच ॥ ३ ॥

विश्वास करो, हे सोम्य! कि जो यह सूक्ष्मता सबका मूल है, यह सब कुछ
इसी से आत्मा वाला है। वह सत्य है, वह आत्मा है, वह तू है, हे श्वेतकेतो!
(पुत्र ने कहा) हे भगवन्! मुझे फिर बताएं।*

पिता ने उत्तर दिया, तथास्तु हे सोम्य।

इमा: सोम्य! नद्य: पुरस्तात् प्राच्य: स्यन्दन्ते, पश्चात् प्रतीच्य:। ता:
समुद्रात् समुद्रमेवापियन्ति समुद्र एव भवन्ति ता यथा तत्र न विदु
रियमहमस्मी यमहमस्मीति ॥ १ ॥

हे सोम्य! यह नदियां पूर्वी (गंगा आदि) पूर्व की तरफ बहती हैं और पश्चिमी
पश्चिम की तरफ बहती हैं। वह समुद्र से समुद्र में लीन होती हैं (अर्थात् किरणों
से पानी समुद्र में से अन्तरिक्ष में खींचा जाता है और फिर मेघ बरसकर बहता
हुआ समुद्र में जा मिलता है) और समुद्र ही हो जाती हैं। वह (नदियां) जैसे वहां
नहीं जानतीं कि मैं यह नदी हूं या वह नदी हूं।

एवमेव खलु सोम्येमा: सर्वा: प्रजा: सत आगम्य न विदु: सत
आगच्छामह इति। तइह व्याघ्रो वा सिंहो वा वृको वा वराहो वा
कीटो वा पतंगो वाद द्ंशे वा मशको वा यद् यद् भवन्ति तदा
भवन्ति ॥ २ ॥

इसी प्रकार हे सोम्य! यह सारी प्रजाएं सत् से आकर नहीं जानतीं कि हम सत्
से आई हैं। वे यहां जो कुछ थे चीते या शेर, भेड़िये व सूअर व कीट और पतंग
व हंस और मच्छर। वही फिर-फिर होते हैं।

य एषोऽणिमैतदात्म्यमिद ꣳ सर्वं तत्सत्य ꣳ स आत्मा तत्त्वम।स
श्वेतकेतो! इति। भूयएव मा भगवान् विज्ञापयत्विति। तथा सोम्येति
होवाच ॥ ३ ॥

जो यह सूक्ष्मता सबका मूल है, यह सब कुछ इसी से आत्मा वाला है। वह
सत्य है। वह आत्मा है। वह तू है, हे श्वेतकेतो! (पुत्र ने कहा) हे भगवन्। मुझे
फिर बताएं।*

पिता ने जवा दिया 'तथास्तु हे सोम्य'।

पुरुषꣳसौम्योपतापिनं ज्ञातयः पर्युपासते 'जानासि मां जानासि मामिति' तस्य यावन्न वाङ्, मनसि संपद्यते, मनः प्राणे, प्राणस्तेजसि तेजः परस्यां देवतायां, तावज्जानाति ॥ १ ॥

हे सौम्य! जब कोई पुरुष बीमार होता है तो उसके संबंधी बांधव उसके पास-पास बैठ जाते हैं (यह कहते हुए) "क्या तुम मुझे जानते हो, क्या तुम मुझे जानते हो?" जब तक उसकी वाणी मन में लीन नहीं होती, मन प्राण में, प्राण तेज में और तेज पर देवता (सत्) में (लीन नहीं होता) तब तक वह जानता है।

अथ यदाऽस्य वाङ्, मनसि संपद्यते, मनः प्राणे, प्राणस्तेजसि, तेजः परस्यां देवतायामथ न जानाति ॥ २ ॥

स जब उसकी वाणी मन में लीन हो जाती है, मन प्राण में और प्राण परादेवता (लीन हो जाता है), तब वह उनको नहीं जानता है।**

स य एषोऽणिमैतदात्म्यमिदꣳ सर्वं तत्सत्यꣳ स आत्मा तत्त्वमसि श्वेतकेतो! इति'। 'भूय एव मा भगवान् विज्ञापयत्विति'। तथा 'सौम्येति' उवाच ॥ ३ ॥

स जो यह सूक्ष्मता (सबका मूल है) यह सब कुछ इसी से आत्मा वाला ह सत्य है। वह तू है, हे श्वेतकेतो!' (पुत्र ने कहा) हे भगवान! मुझे फिर***' पिता ने उत्तर दिया 'तथास्त, हे सौम्य।'।

And finally,

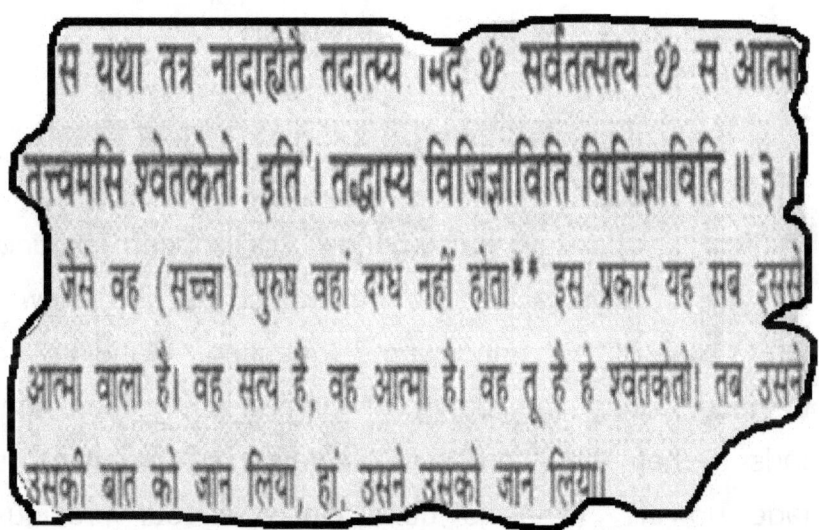

["Chhandyogya Upanishad ", Dynamic Publications India Ltd,
Page-162,160,165 &166]

With this whole exercise, the question arises now is that

"Is it really possible for humans on earth to know what the most fundamental stuff of the creation is? Or,

What the universe is made up of?
Or,

"Is there a limit for human comprehension about the universe?"

Let's say that we flow with modern particle physics first, as we expect that the answer of an unanswered question, standing on our face, since the dawn of human civilization that reads "What the world (or universe or creation) is made up of?" (That is quite closer to our previous curiosity of "Illusion of Solidity") must come from this part of research. And we need to talk about the smallest possible scale in the nature, what is called as "Space-Time Foam" or "Quantum Foam" at the extremes of infinitesimalness of this creation in this universe (or of this universe?) [And I am pretty sure that it is this infinitesimalness that is our one and only hope to find out the answer of our base curiosity and because I do realize that size and even geometry are nothing but the cosmic illusions. As theory of relativity shows in terms of concepts of "Length Contraction", etc that they change if the situation changes and a thing that changes as per the situation, can't be true or real.]

So, let's talk about this "Quantum World" a bit. It was 1955 when John Wheeler proposed this idea, now popularly known as "Quantum Foam" or "Space-Time Foam". It is a topic of Quantum Mechanics, so, it might be little tough. But we have a simpler idea that goes like this.

Imagine somehow, someway, you got "Free Will". So that you could manage to control your own body size (Because as we have just seen that size is a cosmic illusion, so, it won't be hard for you to do so anyway). Like per your wish, your body could be as big as observable universe or it could be as small as a nucleus of an atom or even might be billions of times smaller than a nucleus. [Well, what all I am asking here is just an imagination, like a science fiction and what all you need to do is just sit comfortably, have a coffee or something and just imagine and at the end you will say "Whatever it was, it was pretty awesome".]

So, after acquiring a control on your body size, you leave outer space as it is and as we said that it appears that the very source of this entire physical world of ours here on earth surface is this infinitesimalness or a kind of quantum world and it is this world in which we can find our answers and that is also what modern science is concluding today. So, we decide that you must pay a visit to this quantum world. And you start reducing your body size and now you are as small as an insect. And you keep on reducing your body size. And now, you are as small as a virus. And you keep on reducing your body size. And now, you are as small as a molecule and you keep on reducing your body size. And now, you are as small as an atom. And you continue your journey from earth surface, towards this quantum world (As we did in "White Book") and now you are billions of millions times smaller than an atom. [Is it possible that even if you keep on reducing the size of your body still the journey always remains continuous and never ends towards the infinitesimal nature of this universe? Who knows? But leave it aside, as we need to talk about your present status as an observer in this quantum world.] Do you realize that what exactly you are doing when you are approaching quantum world from the earth surface? Well, you are crossing worlds after worlds, as you did when we were approaching the limits of observable universe starting from the same earth surface. So, now you are as small as in relation to you, an atom is as big as this observable universe, when you were on earth surface. Now, you are in Quantum world. So, what you see here?

As per John and modern science or more precisely quantum mechanics, you will witness a sea of turbulence, a world that is made up of a pure chaos and hence it is far from being calm and so violent that in one instant, one particle (say a virtual particle) comes into existence and in the very next instant it gets annihilated by its anti-particle and goes out of existence. It would appear to you like an ocean of chaos, in which particles are popping in and out of existence that might appear like this.

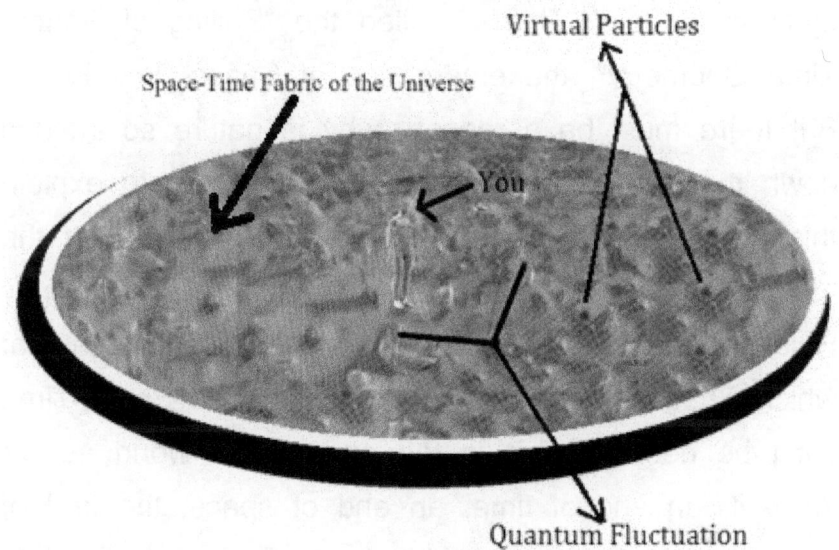

Now, if you can realize that the conventional ideas of what you mean by 'Here' and 'There' and even the ideas of 'When' and 'Then' that give you a sense of what you perceive in terms of 'Space' and 'Time' on earth surface are no longer applicable here and it is this reason that the conventional physics like Newtonian mechanics or even relativistic mechanics is no longer workable here and hence since the beginning of 20th century, scientist throughout the globe are developing a new domain of physics, what we say "Quantum Physics" and the work is still under progress. As per quantum mechanics, Space-Time Foam, if it is really out there (make a note of that too), must be millions of millions times smaller than the nuclear dimensions. The dynamism of the nature is so intense at this scale (10^{-32} cm or Planck scale) that space-time boils. This is called the "Boiling of Space-Time Geometry", the emergence of Space-Time Foam. But there must be a reason, why is nature so intense down there that space -time boils? In order to explain this, we have "Heisenberg Uncertainty Principle" or the "Quantum Mechanical Principle" that was invented by Heisenberg in 1927. It is this virtually infinite scale at which Space-Time boils and beyond which Space-Time can't be defined. This is the end of this world as we know it, an end of time, an end of space, the end of causality. What we should say, a final and absolute limit of everything whatsoever. Now, the bubbles arising from the boiling of Space-Time that is an eternal reality at this place (place?), some of these bubbles grow. They expand exponentially to become universes. As per present calculations,

"Every cubic centimeter of space percolates 10^{144} universes per second."

Now, most of these universes die. They survive for 10^{-44} seconds (Planck Time) and then disappear into their ultimate source. But many of them grow and are propelled to this exponential expansion called "Inflation" and lead ultimately to the emergence of galaxies and clusters of galaxies. So, to conclude, as per this theory,

This is how our universe begins. It is just such a bubble, just an impulse of an unknown manifestation.

This shows the intense creativity of the nature. But what does it implies? The world of ours, here on earth surface that appears so solid and quite obedient for the Laws of Nature (or laws of physics) is coming from an ocean of chaos? As you can realize that this is completely fictitious. But it is exactly what modern science or quantum mechanics says today. Now, despite of its obvious absurdities such as its speculation that "10^{144} universes per second", etc we are spending our time on it because there is something inside this idea that is quite useful for our area of investigation in this book because this foam has been conceptualized as the foundation of "The

Fabric of the Universe". But how a physical existence like you or your human body around you can move, if the universe itself is fabricated? And if it is really fabricated, it shows its perfect resemblance to the idea of "The Matrix". Moreover, this idea of quantum foam can be used as a qualitative description of subatomic space-time turbulence at extremely small distances (on the order of the Planck length, say 1.6×10^{-35} meters). Because the final conclusion of quantum mechanics says that at such small scales of time and space, as we saw, the Heisenberg uncertainty principle allows energy to briefly decay into particles and anti-particles and then annihilate each other without violating physical laws of conservation, like Law of Conservation of Mass and Energy. As the scales of time and space shrinks, the energy of the virtual particles increases. As per "General Relativity", energy curves space-time and causes a distortion in its geometry. This suggests that at sufficiently small scales, the energy of these fluctuations would be large enough to cause significant departures from the smooth space-time seen at larger scales like on earth surface or beyond, giving Space-Time a "Foamy" character. As I said, in relation to other theories, quantum foam is theorized to be the "Fabric of the universe" but can't be observed because it is too small and we don't have that much efficient microscope or any other technology. Also, Quantum Foam is theorized to be created by virtual particles of very high energy.

But how a virtual particle can have energy that is a real stuff?

But virtual particles appear in Quantum Field Theory, arising briefly and then annihilating during particle interactions in such a way that they affect the measured outputs of the interaction, even though the virtual particles are themselves space. These "vacuum fluctuations" affect the properties of the vacuum, giving it a non-zero energy known as "Vacuum Energy", a sort of Zero-Point Energy. This whole stuff is good enough to convince you that it is this knowledge about Quantum world or infinitesimalness of this creation that is of our top priority and if there is any possibility for the existence of "Theory of Everything" (TOE) as it is called (As a matter of fact there is.), this knowledge would be a part of it for sure.

As we saw earlier, it is exactly the same what our ancient literature says, with the words,

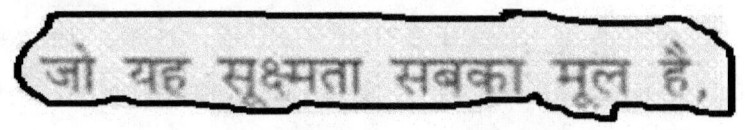

जो यह सूक्ष्मता सबका मूल है,

"..........ye sukshamata jo sabka mool hei."

.............part of the deal.

[I]

After years of investigation, synthesizing and theorizing, now, they come up with their one and only cannon, what is called as "Standard Model" that in order to satisfy human curiosity about this puzzle of "Box within the Box" or what we just saw as cosmic conundrum of "Particle within the Particle" is the most efficient and promising theory. But before the announcement of CERN that they have found what is known today in popular science as "The God Particle", it was presumed that this standard model was incomplete and now they think they have made it. But did it change anything? Do we or they have a new understanding about the nature of the universe now? Because what they say that it is the best possible tool they have invented so far in order to describe the nature of the universe (nature of nature?), so, it will be quite logical to talk about what this standard model got to say about our physical world. Yes? Okay. The Standard Model of particle physics is a theory concerning forces of nature like electromagnetic force, etc, the particles of the universe and the interaction between forces and particles. It was developed throughout the latter half of the 20th century, as a collaborative effort of scientists all around the world. The current formulation was finalized in the mid 1970s upon experimental confirmation of the existence of "Quark" that was proposed by Murray Gell-Mann, for which he was awarded Nobel Prize in 1969. Because of its success in explaining a wide variety of experimental results, the Standard Model is sometimes regarded as a "Theory of Almost Everything" and it says that the universe is made up of only two types of fundamental particles namely "Leptons" and "Bosons". And as you can see, if somehow you get true knowledge of just these two fundamentals of the cosmos, you will get what is known today as "Theory of Everything" (TOE) and now, you can understand entire creation at its very basic level. This theory says, leptons belong to matter and bosons to forces of nature, as they are the presumed carriers of natural forces, from one point to the other point of the same universe (Why they can't go beyond this universe? I don't see any limit or obstacle for them, do you?) Standard model, that is the most successful theory on the subject also contains quarks and leptons as its fundamental particles, with their three generations of each, no one knows why (Make a note of that too as it makes this best scientific tool on the subject 'ad hoc' again) and they say that we do have a "Quantum Field Theory", quite closer to Einsteinian dream of a "Unified Field Theory". *[Well, before we go ahead, we have a story that goes like this. In last 20-30 years of his life, when our A1 used to be a house hold name on the globe, a Nobel laureate and the one and only symbol for human intellect throughout the planet. He was working on a theory, that was so popular with the name as "Unified Field Theory", where he was attempting to combine electromagnetism and gravity, in one unified construct and the potential inherent in his idea, lies in the fact that although he could not complete his dream but it left several mysteries and unsolved question, for instance "Invisibility" and some of them became popular science fictions even today and perhaps it was one true source for the origin of an idea known as "Theory of Everything" (TOE) that says that there is a possibility that all the four fundamental forces of nature viz electromagnetic force, strong and weak force and gravity are just four different manifestations of one "Unified Force" and if we can unify them, then a single theory of this kind will be efficient enough to explain everything whatsoever.]*

Standard model is a theory that can tell us how quarks combine to form protons and neutrons. So that we can learn, how universe was created from its most fundamental level, up to its cosmic limits. So, standard model is the best possible description we have about our understandings of the nature of the universe or creation. But something is weird as our best theory says that on its own, every single particle is massless, like protons or even quarks and all others have absolutely no mass and as relativistic mechanics conclude, only a massless particle like "photon", the particle of light, can travel at 'c' that is the velocity of light. So, if matter particles that make up everything else are massless, what type of nature they can create around us? And in 1964, Peter Higgs came up with an idea that was a theoretical possibility but appeared as very weird solution of this problem. He said what is now known as "Higgs Mechanism", our one and only accepted theory that makes every physical a real physical stuff. Because it talks about how any particle acquires its mass in the universe (and hence the solidity that comes for free with it) and we need to consider what Pete said. The "Higgs Mechanism" is essential to explain the generation mechanism of the property "mass" or how universe becomes massive. And our subject,

How an existence becomes physical in this universe?

As a matter of fact, the mechanism was proposed in 1962 by Philip Warren Anderson but its relativistic model was developed in 1964 by Peter Higgs and he was awarded Nobel Prize for physics in 2013, for the theoretical discovery of a mechanism that contributes to our understanding of the origin of the mass of subatomic particles, and recently it was confirmed through the discovery of its predicted fundamental particle "Higgs Boson" or "God Particle" by CERN's "Large Hadron Collider" (LHC). So, let's see what the Pete's theory suggests to you.

Let's suppose that particles are truly massless. On the very fundamental level, there is no mass at all, in the elementary particles like quarks or leptons. Now, imagine that entire universe is filled with some kind of liquid or presume that in the beginning of the creation you take the universe and fill it with a Quantum Field that acts to slow down the fundamental particles as if this very room is filled with water and if you move your hand through it, it slows your hand down. So, as per this theory, the universe is filled with all pervading, omnipresent "Higgs Field" and because all the fundamental particles keep on interacting with it, everywhere and every time (make a note of that), so, its interaction gives them what is perceived by you as an observer (or more precisely as a perceiver) its "Rest Mass" that on the classical level, creates its solidity for you. Like,

Cause is quantum mechanical and belongs to infinitesimalness of this creation and its effect is manifesting here in our classical world on earth surface.

But if this extraordinary field interacts with every particle, how the differentiation occurs in terms of what we call light particles like photons and neutrinos and heavy particles like Top quarks at infinitesimal level of creation or what you call Quantum world there and even what you feel as light or heavy in your classical world here on earth? Because according to this theory, every individual particle is massless. What Pete suggested is that it basically depends on how much interaction a particle has with this Higgs field. If its interaction is high, it makes it more massive and vice versa. For instance moving a pencil through water is easy than moving a billiard table. Perhaps you said "No Way". "That can't be true". But it is. Pete said so and he won even a Nobel Prize for this and he further says that

"Not only you, but me, this book, this very room, even every goddamn thing in our surroundings, is swimming in this mysterious

omnipresent field, that is present from here up to the far reaches of outer space, towards the infinities of this universe, and hence replacing our old idea of 'Empty Space' everywhere outright."

Well, not only this, he told to BBC (British Broadcasting Corporation) that like photons as packages of energy of some kind of field, the vacuum is the lowest energy state and in it there is a background field that interacts with everything that goes through it, simply because the background affects the wave that propagates through it. And Higgs boson or the God Particle is the excitation of this background field. *[Wait a minute; I tell you something, the idea of 'Space' and its science has a very interesting history.*

In 17th century, when Izz came in, he killed every existing idea about "Space" of his time and invented a new idea of space, what his "Principia" teaches us today in terms of his "Absolute and Relative space". In 1905, when A1 came in, he killed this idea of space completely and invented his new idea of space, what we today call "Minkowaski Space" or simply "Space-Time". And now when Pete came in, his theory implies that we need to rethink what we mean by the term "space" again. So, you see this apparently simple idea of space is far from being simple and there is still a huge potential in the human curiosity that "What is space?" Scenario is like, if you go to Aristotle with the question "What is space?" he will teach you something about space and if you go to Isaac Newton for the same reason, he would say that wash away all the things Ari has told you and he will teach you something about space. And now, if you go to Albert Einstein with the same question, he would say that whatever Izz and Ari taught you is not complete and he will teach you something else. But if you go to Peter Higgs with the same curiosity, he would say that whatever Ari, Izz and A1 had taught you is not true and now he will tell you something else. I mean to say, how many times we need to learn about something like space? Not only this but it is quite probable that a future scientist will come (or has come?) and he will tell us that whatever Ari, Izz, A1 and Pete told us about space is not the complete truth and he will teach us something else about space.

Well, what it means to you is that despite of an impressive advancement of technology, do we really have an answer of a simple question that "What is Space?" What this is that exists everywhere, starting from the very surface of your eyes up to the far reaches of outer universe? What this is that is the fundamental domain of "The Matrix" itself? And what this one of the five most fundamental elements (viz Space, Fire, Air, Water and Earth) of this creation, really is? Because if space remains unknown and mysterious then how the truth of any existence or any event, which occurs inside of space could be known?]

Well, you would say that if this is an omnipresent field and gives every physical its mass, even if I am not, but my body that is a human body is a physical stuff. So, it must get its mass too from this field. So, why I don't feel it? (Make a note of that. It is really a good question and we will need it in Blue Book) A bizarre response for your question is that it is the very interaction between the particles of matter of your body and Higgs field at quantum level due to which you experience your own body, for example its inertial nature, when a force is applied on you.

Now, there is an interaction between you and this Higgs field that creates a human body around you and as a consequence, you realize that you have two hands, two legs, a nose, and two eyes and so on. It is this field that creates the complete existence of an observer like you and me. And perhaps this is the reason, why observer itself can't perceive it. The very source of you, as an observer is this omnipresent Higgs field and as Law of Nature of "Cause - Effect Relativity" says, an effect can't go beyond its own cause.

And the intensity of this natural conclusion of the Higgs mechanism has several surprising consequences such as it could lead to the very origin of the observer itself, even the origin of human experiences combined with human senses themselves, like touch and vision.

Well, the idea of the "Field" is always tricky since Mike invented it, so, the point that "can you feel a field?" is a bit tough. You see it's always here, like earth's magnetic field for instance, in general you don't feel it but if you have a campus, you can feel its presence right here. Same is the case with electromagnetic field, as you don't feel it but when you tune your radio, you find it's here. Similarly, this omnipresent Higgs field is here with you but you just don't have a device or something advanced enough to feel its presence. But is it just a matter of technology to invent a device to detect this Higgs field? Look at this, there is a huge difference between other conventional fields and this Higgs field as if we turn off the dynamo mechanism at earth centre, the magnetic field gets disappeared and in general if you turn off the source of any field, it gets disappeared completely. But Higgs doesn't have its known source, so you just can't turn it off, it is always here and there and if you ask Pete why is that, he would say "I don't know but it is there in the very emptiness of this universe since its birth" (Like a background of the universe?). Make a note of it as field is always an effect of its cause that is its source and "Cause - Effect Relativity" is a firm Law of Nature but apparently Higgs does not obey this law of nature.]

If you can imagine, something like a science fiction and say you invent a device like a remote, quite similar to a remote of your TV set and you point it towards something and with a "Click", you are able to switch off the Higgs field at that point, whatever is there will just discombobulate because it will become massless instantaneously and it will start moving at 'c', the velocity of light and it will go away from you. Or, if you could control the strength of Higgs field by your remote, you could make lighter things heavier and heavy things light, in whatever way you like (mountains into sand grains and sand grains into mountains). Because now,

The power inherent in the idea of mass and inertness of a physical existence in this universe will all be yours.

So, modern science concludes that what we mean today by the term vacuum or void or nothingness or emptiness in the universe is if you remove all the matter particles from a void and you turn off all the other fields, still the empty space would contain this omnipresent Higgs field (everywhere, all the time). And it will give you another science fiction movie or a book because the idea of mass is intrinsically connected to the idea of a chemical reaction. So, a slightest change in Higgs field strength would give you a brand new chemistry, far different from the chemistry of our known 98 elements of modern periodic table. And you would enjoy doing this, if you want to visit new and completely different worlds. For example,

You visit a world in which the nature of Higgs field is quite different from our world and in that world you find an element that doesn't exists here in our periodic table but that is far more radioactive than even Radium and other elements and when you come back, you realize that you don't need an Atom Bomb anymore. Now, you have far more superior power in your hands.

So, there is a "Field" (Higgs Field) and there is a "Fielder" (You) and a unification of these two existences creates new worlds.

But the question is "Can we do it?" Well, no one knows even the fact that is it possible or just a theoretical possibility. Moreover, we do know or we think we do know something about it. It is believed that at the time of birth of the universe, what we call the "Moment of creation" in terms of Big Bang, just after the birth of the universe, this Higgs field was turned "OFF" and in those days (or in those picoseconds to be precise), the universe was a very different place. There was not even any single atom or anything in the universe. Every particle was moving at velocity of light, so, you could not build any physical structure there. And after this era, Higgs field was turned "ON" and things were getting slower and then there was a possibility that they could come together to create new stuff like stars and galaxies and this is continuous till date and rest as you know is just a consequence in terms of the time dependent evolution of the universe and of humankind inside of it. But nobody knows, what was the cause behind this control, who turned Higgs 'Off' and then 'On'? God did it? Well, as you say.

[J]

B ecause idea of inertia is intrinsically linked to an idea of mass and because origin of mass in this universe is as mysterious, as it used to be in its ancient times. So, there is a possibility that in order to understand the truth of inertia that is our subject here, we might get some clues from this point as well. So, let's talk about it for a while. As the text books on the subject show, "The Big Bang theory" (the "Inflationary Big Bang Theory" to be precise) is the prevailing cosmological model for the origin and the early development of the universe. The key idea is that

Our universe is expanding that means the universe was denser and hotter in the past (or in its past). So, the Big Bang model suggests that at some moment, all of space was contained in a single infinitesimal point that is considered as the beginning of the universe. Modern measurements place this moment at approximately 13.8 billion years ago. It is considered as the "Age of the Universe" itself. After the initial expansion, the universe cooled down and hence allowed the formation of subatomic particles, including protons, neutrons, and electrons. The majority of atoms produced by the Big Bang were hydrogen, along with helium and traces of lithium. Giant clouds of these primordial elements later coalesced through gravity to form stars and galaxies and the heavier elements were synthesized either within stars or during supernovae and this whole cosmic operation is continuous till date.

As you can see in terms of "Expansion Universal", the distance between the galaxies increases today, this implies that in the past, galaxies were closer together. So, the known laws of nature can be used to calculate the characteristics of the universe in detail back in time to extreme densities and temperatures. While large particle accelerator can create such conditions because as we saw in "Green Book" that if you take two particles, like protons and anti-protons in a particle accelerator and if they collide, they create a "Fire ball", which is quite similar to the Big Bang and the only difference between these two events is of scale, resulting in confirmation and refinement of the details of the Big Bang model.

George in 1927, proposed the idea of Big Bang. But our A1 remained at focus point of this whole development. Because it was his theory of general relativity that became the foundation of what we today call modern "Big Bang Cosmology". In 1929, Ed discovered that the distances to far away galaxies were increasing with time. His observation was taken to indicate that all distant galaxies and clusters have an apparent velocity, directly away from our vantage point with just few exceptions that is the farther away they are from us, the higher the apparent velocity of them, regardless of direction (make a note of that). Presuming that we are not at the centre point of a giant explosion, the only remaining interpretation is that all observable regions of the universe are receding from each other in a time dependent manner.

In the mid 20th century, scientific community on the globe was divided between supporters of two different expanding universe theories namely the "Big Bang Theory" and the "Steady State theory". Observational confirmation of the Big Bang scenario came with the discovery of the cosmic microwave background (CMB) radiation in 1964, and later when its spectrum I.e. the amount of radiation measured at each wavelength and the results obtained from "W-

MAP" (Wilkinson Microwave Anisotropy Probe, 2001) and "COBE" (Cosmic Background Explorer, 1989) satellites were found to match the thermal radiations of a black body. Since then, astrophysicists have incorporated observational and theoretical additions into the Big Bang model and this whole work just killed the steady state theory outright.

Keep "Big Bang" aside; we really don't need it here. I told you about Big Bang just because I had to and if I am promoting this idea, I am sorry for that and I am sure you know what I mean. You see, I said that I will give you a clear picture about our subject, so, whatever appears linked to the subject, I want to talk about. Well, what I think? Big Bang as a theory or I should say a science of today because kids (may be yours) all around the world are learning that as per modern science, which is like "Bhagawat Gita" or "Bible" "Koran" for them says that the origin of the universe is from Big Bang and not only this, despite of it's all absurdities (For instance by using this model they can speculate about what happened one billionth of a second after the Big Bang, 13.7 billion years ago. But they can't tell you what happens even after one moment when a chemical reaction starts right here, right now). The idea of Big Bang is now, a part of the text books on the subject throughout the globe and even students of 10+2 level are studying it as the truth about the origin of this universe. Why is that? Because the top notch human intellectuals of today are promoting it as it is a fact. But my conscience says that we still don't know how and why universe was born. Sometimes I wonder how they can do that. They just don't want to see it or what because it's simple. I tell you how. Just erase all the theoretical part from the idea of Big Bang, as if nobody ever taught you what is Big Bang and you never had any idea, how the universe was born. Like a complete brain wash on the subject. Now just collect all the astronomical observations and experimental results, we have today about the outer space and start thinking all over again. You will certainly realize that Big Bang theory is full of drawbacks, if you don't want to say, it's a complete bullshit. [For instance idea of Big Bang can never tell you how and why our home galaxy Milky Way acquired such a beautiful appearance or a brilliant optical phenomenon or simply its geometry. Well yes, of course, they can explain this too but it is just a different issue.] But in order to do so, perhaps the only obstacle you will face is the observation that the universe is expanding with an accelerating pace. But I am sure; there is logic for that as well, an unknown truth.

Well, we need to talk about what is symmetry? Symmetry is a formless union of matter and anti-matter. So, when the universe comes into existence, there must be a transformation of supersymmetry (SUSY) into a Broken-supersymmetry (B-SUSY) like water goes through a process known as "Phase Transition", when it gets converted into ice.

Now, if we want to invent our own model of this universe (Make a note of that too), then our new model or a new "Cosmological Model" says that

The birth of our universe has two completely different aspects first, "An Origin of an Absolute or Improper or Actual universe". And second, "An Origin of a Relative or a Relativistic or a Proper or a Potential Universe".

Now, the first aspect or the "Birth of Absolute/Improper/Actual Universe" has approximately nothing to do with human understandings. [See, it is exactly the same universe, about which our A1 said "The most incomprehensible thing about this universe is that, it is comprehensible."]

But, the second aspect of this absolute conundrum, namely "The Birth of Relative or Relativistic or Proper or Potential Universe" is definitely a subject matter of human endeavour. And basically this birth is nothing but the origin of our home galaxy "Milky Way", in its absolute domain that is an absolute or actual universe to which we have seen it, in form of infinite darkness, since the beginning of this book, where you made a note of this and when we were talking about your existence in an inky void in the "White Book".

So, the best possible way in order to continue our journey towards the absolute truth is that we can presume that galaxy or our home "Milky Way Galaxy" is nothing but a quantum of an "Absolute Field" or the quantum of the fabric of the Absolute or Actual Universe. And what is even more interesting about this is the fact that because like a law of nature, every physical stuff whatsoever is quantized, by its fundamental or quantum level. So, if you invent the truth of a single quantum of any particular physical quantity (And here the quantity is of course our universe itself), and just by doing calculations or a sort of calculus, you can find out the whole truth, about that physical quantity.

Similarly if you presume that milky way is really a relative or relativistic or potential universe or if it really is a quantum of the cosmos then it is for sure that you can create a new cosmological model that would be approximately 170 billion times more focused and efficient then the cosmological models that were taught to us, by science and technology of today.

Having said that now, we really need to get back on track, where we are talking about "Symmetry". So, it appears that

This infinite darkness that is undoubtedly ubiquitous in nature is nothing but a "Supersymmetry" (SUSY). And this SUSY in its absolute domain that is this absolute or actual universe is basically "A State of An Equilibrium" of the universe.

At the moment of creation (13.7 billion years ago?) there was an "Absolute Phase Transition". When universal equilibrium got broken down and supersymmetry (SUSY) was transformed into a broken supersymmetry (B-SUSY).

But here comes a question. What is this Absolute Phase Transition? Basically, in nature the best possible example of it would be the transformation of water into ice (as it is shown in the picture below), which is just a localized phenomenon. But this absolute phase transition, we are talking about, is a sort of cosmic event.

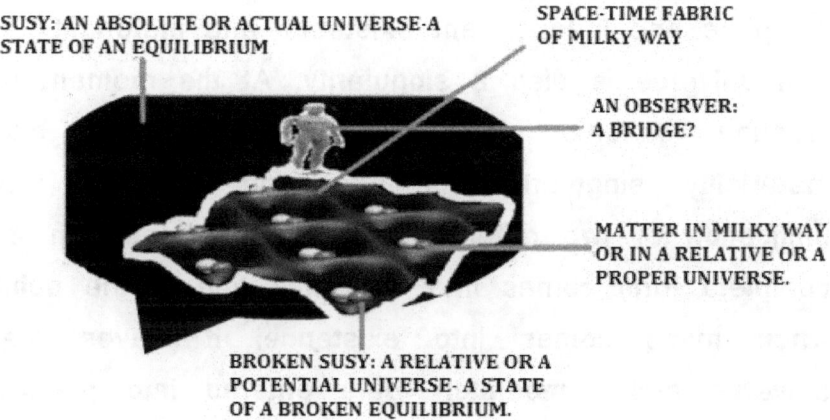

SUSY: AN ABSOLUTE OR ACTUAL UNIVERSE-A STATE OF AN EQUILIBRIUM

SPACE-TIME FABRIC OF MILKY WAY

AN OBSERVER: A BRIDGE?

MATTER IN MILKY WAY OR IN A RELATIVE OR A PROPER UNIVERSE.

BROKEN SUSY: A RELATIVE OR A POTENTIAL UNIVERSE- A STATE OF A BROKEN EQUILIBRIUM.

You know what. I got something to tell you, this might be a bit confusing, as it is exactly opposite to our human common sense. And it is totally up to you that how much seriously you would like to take this idea because I won't be able to provide you any proof on this and you can even ignore it completely, if you want to.

Real universe is a singularity. Matter as a physical quantity, when exists in real universe is a complete and independent existence and matter in the real universe is a singularity. On the other hand, materiality that also exists in real universe is a complete and independent existence and materiality in real universe is also a singularity. At the moment of creation, a union between matter singularity and materiality singularity happens and due to this unification of two different singularities, "Mass" in its complete form comes into existence and at the point when mass comes into existence, it leaves real universe and immediately gets entered into pseudo universe to which we recognize it as "Creation". So, the process happens into real universe or "REALVERSE" and the effects come into pseudo universe or "PSEUDOVERSE" or "Creation". Or, the cause exists into real universe or Realverse and effect exists into pseudo universe or Pseudoverse or creation. [And this is the reason, why even after the years of research and global efforts, still we couldn't find the true cause of this effect, we call our "Physical World".] Now, at the point on which mass comes into existence by the unification of these two singularities, it originates a duality to which we recognize it as "Wave Particle Duality" and this duality is nothing but creation itself. So, origin of creation is from wave particle duality and it originates entire creation, as a modified version of "Wave Particle Dualism". So, this fundamental concept of quantum mechanics called as "Wave Particle Duality" and this present creation are two different aspects of a single reality like two different manifestations of an unknown. On the quantum level, creation is nothing but a wave particle duality and on macroscopic level, wave particle duality is nothing but creation itself. Quantum mechanics concludes that if you observe an electron, it would behave like a particle. But if you are not looking at it, it would behave like a wave. So, you can see that the concept of "Wave-Particle Dualism" is quite puzzling but what it can tell you is the fact that it is the observer that is you, who is the most fundamental stuff in the whole subject. But since the dawn of modern science from Aristotelian Mechanics to Newtonian or Classical Mechanics to Relativistic Mechanics and finally to Quantum Mechanics of today, the true essence of the idea of "An Observer" is still missing. "If there is a beginning, there is an end". True, but it is incomplete because "If there is no beginning, there is no end". The problem is that the first half of this law of nature looks scientific but the second half looks like fiction. At the moment of subtraction that is the opposite of creation, the reverse of this happens. Matter-materiality combination breaks down and as law of nature says that "Nothing can exist as a singularity in the creation", so, matter and materiality both simultaneously loses their existence from this "Pseudoverse" or pseudo universe or this present creation and immediately get entered into "Realverse" or real universe or their final destination. So, this question, "Where does the mass come from in the inflation phase of the Big Bang, which makes an ultimate source of creation and where does it go when it disappears at the centre of the Black Hole?", doesn't exist in its absolute sense but of course there is no denying that in a relativistic universe it exists because in an Absolute Frame,

"Nothing is going "In" and nothing is going "Out."

If it appears that these two fundamentals, first one is universe and second one is Creation/Physical World/Nature are complementary to each other, with the appearance that the universe is in creation or the creation is in the universe, it seems possible that these two apparently complementary concepts, first one is universe and second one is creation are not at all complementary to each other and neither universe is in creation, nor creation is in the universe. It is the "Human Brain" as a reference frame that connects these two concepts to each other. And if you leave "Human: A Frame of reference", universe and creation are not at all complementary to each other and neither universe is in creation nor creation is in the universe. Now, the question is what is "Existence" exactly? Whenever you observe an entity/existence (Do you remember that just few moments ago, you picked up a stone and you were looking at it, as you never saw such a thing in your whole life before?), you feel that it is an existence like yourself and any other existences of this universe but this observation of an observer that is originated by human common sense is virtual.

Every existence of this present creation of this universe or in this universe (whatever you assume) is not at all an existence and in fact, it is good to say that every existence that belongs to this creation is a "Resultant", a resultant of the combination of different existences. At the moment of creation of an existence, a certain "Union" (Yoga) happens between different existences and due to this unification, a universal existence comes into existence with respect to this present creation and with respect to every other existence of the same creation.

Look, the number system that is taught to us since the beginning of our education, is purely a man made stuff because universe doesn't work on these "Numbers". But undoubtedly there is a strong appearance that it does. And this is the reason, why "Cosmic Mathematics" is still beyond human understandings and why it looks chaotic at its first place. Well, it's unbelievable. But there is a chaos, everywhere in the "Calculus" itself and hence in the whole body of "Modern Mathematics". And in the cosmic mathematics and as far as workings of "The Matrix" is concerned, one plus one could never be equal to two, as "Somethingness" can never originate somethingness itself. And of course it is also true that "Nothingness" also can't originate nothingness itself. It is the same case as "Several stones if combine together, can never create a house."

If an observer is in the ship and the ship is on a certain uniform velocity then the observer who is present inside the ship, can never observer the velocity of this ship and if he leaves the ship and comes out of it then he can observer that the ship was on that velocity. Now, the ship is this creation and you are the observer, so, from inside of this creation, you can never conclude that this ship called as "creation or physical world" is on at rest or on a certain velocity and if you go beyond the creation (Not beyond the earth) then only you can find out what the reality is.

Well, this idea has enough potential to show you that our idea of absolute motion is solid. And I am pretty sure that you perfectly understand that it is exactly the same truth, when you are in your inky void and wondering where I am.

[If you go and ask to Ari, he would say that the universe (the entirety of somethingness) came into existence from nothing or nothingness and even modern science apparently also supports this Aristotelian assumption today but just leave it for now.]

YELLOW BOOK

[K]

Now, the obvious question you might ask that this Higgs field, the God particle (Higgs Boson), the Big Bang,....blah blah blah, is this all true? Well, I don't know, nobody knows in fact but what I can tell you is that it is exactly the same point on which the most expensive research is going on the globe now a day as we saw in "Green Book" about CERN (European Organization for Nuclear Research) and its LHC (Large Hadron Collider) that they have made a huge particle accelerator that is a billion of dollars project and this machine is the most expensive and big machine ever made in the history of human civilization. What it all means to you is that it really deserves your attention.

As we have seen earlier that they are creating the collisions of the fundamental particles like protons head on, so that they can create a tiny "Ripples in Space-Time" that could be the Higgs bosons, a proposed field particle of this Higgs field and because of its immense potential in terms of the creation of the universe, it is called as "God Particle" *[Originally its name was "Goddamn particle" because it is quite hard to catch it in a particle accelerator that in time and with the blesses of international media, became "God Particle"]*. As this room is filled with electromagnetic field and hence filled with photons as its field particles. So, if Higgs field is something real, it must contain its particle that is Higgs Boson and if we could create and detect this particle in a particle accelerator, we could be sure that our best possible description about the truth of the universe and about its nature is really true and not just a mathematical jargon. And the good news for you is, on July 4, 2012, they announced that they have found it that is we have discovered this "God Particle".

But as I said, did it change anything? Do we now have a true approach towards the truth? Well, yes, it changed something or I should say everything, as before it, nobody knew who Pete is but he is now a brand name on the globe and a Nobel laureate. But did it change anything else? Even if you keep your faith in what they suggest to you and presume that everything is right, there is an omnipresent Higgs field here, there and everywhere (like God? Because omnipresence used to be the property of God since the ancient times), there is a serious problem with it. It has been known for decades that if there is a Vacuum Energy in the space that is these Higgs bosons, it must have weird impacts on the universe as a whole. Einstenian general relativity has shown that there are only two possibilities, universe either must expand or it must contract *[Quite interestingly the inventor of the theory himself concluded that universe is static that the universe is neither expanding nor contracting. In 1929, Ed could tell our A1 about the expansion of the universe and the scenario could be entirely different. But just leave it aside, happened happened anyway, who cares except A1?]* And it is certain that universe is expanding today.

Take a box like region of universe and fill it with matter particles and now let it expand. The number of particles will remain the same. But if you take the same piece of universe and fill it with vacuum, which it naturally is and then let it expand, the amount of vacuum must increase and hence there will be a consequential increase in the vacuum energy too. So, you will get more and more energy, as the universe keeps on expanding that will increase its expansion rate even more and more and that means it is happening since the time of Big Bang, approximately 13.7 Billion years ago.

So, at the end, you will get an inescapable conclusion that the universe will go a runaway accelerating exponential expansion. Standard model says that the Higgs field is present everywhere and it even calculates that the Higgs field in one cubic centimeter is more than the trillion tons. But if you put this number into Einstenian equations, it comes up with the result, which says that

In less than a nanosecond (or 10^{-9} seconds) this universe expands a billion times.

And that means something is terribly wrong with our present understanding about the subject.

Today, we have a modern science that has an extraordinary power of explaining everything whatsoever. In other world, we are living in a scientific era where anything can be theorized, no matter either it makes sense or it is just bullshit. And here is a theoretical solution to this problem too.

See, Higgs field really is filled in empty space with more than a trillion tons per cubic centimeter but let's say there is one more field, an ad hoc field that was never a part of science before and invented just for this purpose only. And it contributes all of the space with more than trillion tons per cubic centimeters but in negative and hence cancels out this whole.

In general, we don't have negative energy as everything we have around us is a positive energy. But quantum field theory allows vacuum to have a negative energy and the conclusion is perfectly mathematically consistent with the theory. Now, the possible theoretical solution to this problem is that this cancellation has been already happened in the past of the universe.

Tell me one thing. When you look into outer space through your eyes or with a telescope, do you ever realize that what you see is the universe, as it existed in its past? Because, say if you observe a galaxy through your telescope that is millions of light years away, it simply means that the light you receive from it now, in your telescope and then on your retina and then you get an image of the galaxy in the back of your brain, left this galaxy millions of years ago, which simply means that you are observing this galaxy, as it used to be in its past, millions of years ago and it is quite possible that in your present here on earth, this galaxy doesn't exists there anymore. As we saw earlier, even when you observe the Sun itself, you always look at it as it was approximately 8 minute ago in its past. Because light you receive from it, that gives you its observation, left the Sun 8 minutes ago. (If the Sun explodes in the future and if you are still on Earth, you have 8 minutes to do, whatever you can do. Yes?).

So, if you observe a galaxy billions of light years away, you can have a window to look back into the past of the universe, as it used to be like, billions of years ago. Now, if you observe distant Quasars or Supernova that is a burst of a star when a star dies, its brightness is equal to billions of Sun like stars and hence it gives an explosion of light, quite similar to a little Big Bang and you observe it here on Earth and you can calculate the distance between you and that part of the universe and by calculating few more you can find out how fast they are moving away from us and hence you can measure the rate of expansion of the universe as a function of time and this is exactly what they are doing since 1929, when the discovery of expansion of the universe originally happened in science. What did they find? Well, they found something that was a shock for everyone as it is exactly opposite to what our description about the universe says. We have two points here; in 1929, it was found that universe is expanding that was a perfect shock for the scientific community of that time. But in 1998, it was found that universe is not only expanding, but it is expanding with an accelerating pace, another shock for the science of today. And in the past universe was expanding slowly that means it's this vacuum energy that is driving the universe in its own way. So, we are actually going through an exponential expansion phase of the universe and we are in it right now. And that is shocking. Because the final conclusion of this whole is that

We are living in a universe that is expanding probably even faster than the velocity of light.

So, this vacuum energy is the dominant substance in this universe, as it drives the universe itself and hence determines our fate. Worse still, we don't know any goddamn thing, what the hell it is.

[L]

We have said a lot that we need to reinvestigate the subject that is "Inertia", as we have already seen a strong probability that there might be an error within the subject. So, as we have decided that if we really want to approach towards the truth that "what is inertia?" we need to consider every single point that might have a link with our subject. Okay, so, here we go again.

As it is obvious that everything needs a domain for its existence and our subject that is inertia must also have its domain. And as everything has its origin from the birth of the universe, what science says the "Big Bang". We saw earlier that the universe in its entirety is expanding quite similar to an inflating balloon and getting bigger and bigger in a time dependent manner. So, if it is true then it is quite easy for you to understand that it must be smaller in its past, as somehow you stop the cosmic time and start it running but now in backward direction. Everything that is apparently moving away from each other today will start coming closer and closer and closer and closer and everything will meet on a single cosmic infinitesimal point. But this point is far different then our idea of a point, what Sir Stephan William Hawking described it as "Big Bang Singularity", as if you keep running time backward for approximately 13.7 billion years, that is the estimated age of the universe itself, everything will come together on a singular point. Now stop time and let it run forward. What will happen? You will observe that the universe bursts in order to come to existence; it is the start of everything, what we call "Moment of Creation". But as there was no space before this burst, so, where this cosmic fire ball came into existence? Well, it happened in a total darkness and universe came into being out of this infinite darkness, as even light didn't exist as well.

[Hey, you know what. It is quite interesting that there has been a lot of research on the idea of light and we have a "Science of Light" today. But since the beginning till date there is no real work done on this aspect of the same universe, what we say "Darkness"

(at least not in a direct context) but importance of this idea of "Dark" or "Darkness" lies in the fact that it is a direct and one and only relative counterpart of light itself, whose science as we have seen in terms of evolution of modern science, even since the time when Izz was working in the field of "Optics", plays a crucial role in our modern science. So, there must be a perfect yet unknown "Science of Darkness" that is equally important for us as it is the case with the idea and science of light.]

But as an observer, you have to observe all of this from inside because there was no outside, as space simply didn't exist either. So, the only option left for you is that you have to enter into this singular point in order to find the ultimate answer of our base curiosity.

[I am pretty sure, you can see right here that if you take this "Singularity" either like a "Big Bang Singularity" or like a "Black Hole Singularity" (and even "Mathematical Singularity" as well) at its face value, it is quite obvious from this whole that the very source of everything is this singularity itself. And even the most fundamental stuff of this universe like space and time and hence matter (and mass in it) has its origin from this singularity. In mathematics, a "Singularity" or a "Singular Point" is in general, a point at which a given mathematical object is not defined. Or it's a point of an exceptional set, where it fails to be well-behaved in some particular way. And our point that demands your consideration is the fact that you or me or any other natural existence, everything is exactly the same mathematical object that is described in this definition of a "Mathematical Singularity".]

And as a consequence, all these apparently different singularities or singular points viz Black Hole Singularity, Big Bang Singularity, Mathematical Singularity and any other sort of singularity of this creation, they all are different manifestations of a single and yet unknown reality. So, a final conclusion that can be drawn from this is that

An absolute source for the entire creation is this "Singularity".

And if you can recall, it is this singularity that we have discussed, when we were talking about the "Realverse" or "Real Universe" and "Pseudoverse" or

"Pseudo Universe" or "Creation" in the Green Book. Not only this, it also gives us at least one true "Absolute" that is an "Absolute Source of Creation".]

So, it is quite obvious that the ultimate source for the answers of our curiosity about this physical world is this singularity that is quite similar to a black hole singularity as it exists at the Active Galactic Nucleus (AGN) of our Milky Way galaxy and as we have seen earlier that the modern observational astronomy concludes that most galaxies contain a black hole or this singularity at their centers. *[Make a note of this as it is really important. For instance, why there is a singularity at the heart of approximately every individual galaxy of this universe? And if there was a Big Bang, how universe managed to give one-one black hole to every single galaxy? By gravitational collapse? I am not satisfied. Do you?]*

Hence from a singular infinitesimal point of infinite density, the universe came into existence, like a burst of a fire ball and then it started expanding and till date it is expanding and will continue to do so, in its future as well. So, in this violent beginning, everything like space and time, matter and energy was burst into existence.

And here comes another titan (Titan? Yes, because I don't think that it really belongs to the first order of natural powers). If an accelerating expansion of the universe is a cosmic operation, who is the operator? And here comes the "Dark Energy". The dark energy is a hypothetical form of energy (Hypothetical in the sense that not even single experiment so far has confirmed its existence.) but it is believed that it permeates all space and tends to accelerate the expansion of the universe or if universe ever realized that it needs an operator for its expansion, the dark energy could be its first choice. Dark energy is the most accepted hypothesis to explain the observations since the 1990s (1998 to be precise), indicating that the universe is expanding at an accelerating rate.

According to the standard model of cosmology, the observable universe contains 26.8% dark matter, 68.3% dark energy (for a total of 95.1%) and 4.9% ordinary matter. So, you can see that the dominating stuff of the universe is this dark energy. Dark energy has been used as a crucial ingredient in a recent attempt to formulate a cyclic model of the universe. But as I said it is still "Dark".

So, the scenario at present is very interesting.

Particle experimentalists with their particle accelerators can find out the truth of Higgs and on the other hand astronomers with their telescopes can find out the truth of the expansion of the universe, if it is changing with time or not and combined together it will lead us to a true understanding of the empty space "The Science of Nothing / Nothingness / Emptiness" (That in turn might give us a "Science of Darkness").

As we saw, in previous few decades, the most popular domain of science and of course, the most expensive research, what we just called "Big Science" were particle accelerators, with a quest to find, what is known today as "The God Particle" and as we have seen that there is an intrinsic link between the area of our investigation in this book and this idea of God particle, so, let's talk about it a bit more. Scientists or we should say an army of scientists, worldwide is working in the field of particle physics in order to complete their theory called as "Standard Model", they are looking for a missing piece, what they presume that it exists but they just never found it.

So, they decided to invest billions of Euros to construct a particle accelerator as it is working in CERN, as we saw in Green Book and the purpose is to create high energy particle collision, so that they produce smaller particles to tell us, what they are made up of and the Higgs boson or our God particle, that their theory says is responsible for the acquisition of mass within the particles. Moreover, as we have discussed, the theory goes like this.

"There is an all pervading Higgs field (Without a known source? Well, yes) and it is everywhere (Like God? Yes)."

When a particle like a proton or a quark interacts with this Higgs field, the particles of this omnipresent field, known as Higgs bosons or our God particle causes a drag on this particle and it is this interaction or drag that gives every particle like protons and electrons, its rest mass (and now it can acquire its inertial nature). What the theory suggests is that on its own every single particle is massless. When it interacts with this Higgs field, it acquires its rest mass as per the quantitative interaction of it with this field.

Now, because everything is made up of particles and as per the theory every particle on its own is massless. So, when a physical interacts or its constituent particles interact with this field, it becomes massive (and inertial too). In a simplest possible way, the point is that there is a field everywhere in space like a magnetic field in the surroundings of a magnet.

Every physical object that is made up of particles, interacts with this field, everywhere and all the time and it is this continuum of interaction between a physical and this field that creates rest mass of that physical, due to which it looks massive and hence shows property of inertia, when other things like a classical force interact with it.

And this is the reason, why our A1 said that

"It appears that inertia originates, in some kind of interaction."

And a thing that originates in a mysterious interaction can't be a fundamental stuff of nature but must be a highly mysterious aspect of this universe and it is our endeavour here to find out, what this cosmic mystery is. Now, I tell you another thing that will help us to formulate our new understanding about inertia and the nature of this physical universe.

There used to be an interesting theory within the scientific realm but it is not alive today. But because of its promising indication that it might be helpful for our present investigation, so, let's talks about it a little.

[M]

I t was late 19th century, "Luminiferous Aether", meaning light-bearing aether, was the postulated medium for the propagation of light (Well, today it has been replaced in modern physics by relativity and quantum theory that is a different issue).

Because James firmly established the fact that light is an electromagnetic wave that travels with a very fast but at a finite speed, what we today say as 'c', the velocity of light and hence there must be a medium, so that this wave could traverse from one part of the universe to the other part of the universe (Of course, not between two universes but why not?)

We need a picture in order to make it clear that how could you as an observer travel if this aether is really out there, in the empty space. And here it is.

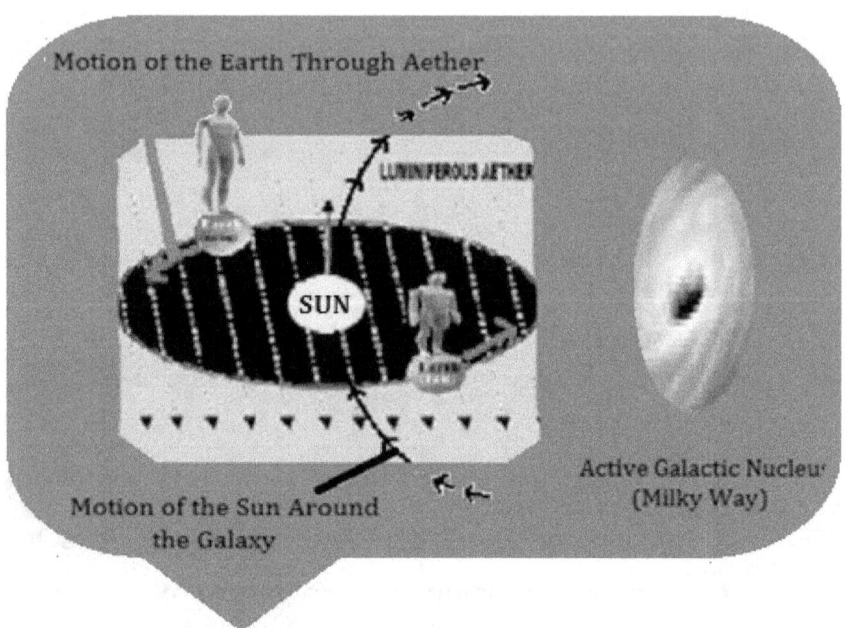

Motion of the Earth Through Aether

LUMINIFEROUS AETHER

SUN

Motion of the Sun Around the Galaxy

Active Galactic Nucleus (Milky Way)

As we said, James was done with his brilliant establishment of the science of electromagnetism and he was one of them, who really invented a new "Science of Light", when he proved that light is a wave that has a finite velocity and almost everyone was sure that if it is the case, like every other wave, light must propagate through a medium. But due to its mysterious nature, nobody had a clue that what it could be. So, they all believed in the existence of an unknown and highly mysterious medium out there in the empty space, what they called "Aether" and it was presumed to be the medium, so that light could travel from one point to another, in the universe that is the motion of light in the creation is through this unknown substance "Aether". Everything was fine until "Annus Mirabilis" (or Miraculous Year) or "Year 1905", when A1 came in, out of the blue *[Well, he really did come out of the blue, as it is just not normal that an unknown person and clerk of third grade in an office, could come up with some scientific ideas to rule rest of the scientific community on the globe for eternity.]* and backed upon the outstanding results of Michelson & Morley Experiment (1887), he was one of the ones, who just killed this whole concept of Aether and as a result, it was replaced by his theory of relativity and Quantum Mechanics. But we need to talk about it as it is quite closer to our area of investigation here. Even A1 in his later life also realized that this concept of "Aether" was not a complete bullshit after all. Well, what is in it of our interest? Let's find out. Izz once said.

"I don't know what this Aether is, but that if it consists of particles then they must be exceedingly smaller than those of Air, or even than those of Light. The exceeding smallness of its particles may

contribute to the greatness of the force by which those particles may recede from one another, and thereby make that medium exceedingly more rare and elastic than air, and by consequence exceedingly less able to resist the motions of projectiles, and exceedingly more able to press upon gross bodies, by endeavoring to expand itself."

Now see, particle experimentalists at CERN have claimed that they have found Higgs bosons as we saw earlier that means there must be a field, what they called Higgs field and everything whatsoever, travels through it and hence light also travels through this Higgs field but the photon, the presumed particle of light, just doesn't interact with it and that's the reason it is massless and hence photon is allowed by relativity to travel on 'c' that is the velocity of light.

And perhaps it's the sole reason, why velocity of light is a universal constant. As you can see, there is an obvious overlapping between the idea of Aether and the idea of Higgs field but not just in a simple way but we need to find out a way and we will for sure.

[N]

We have already seen that it was 1929, when Ed (Edwin Hubble, USA) discovered it and as a consequence today it is a well known fact that universe is expanding, galaxies like our home galaxy Milky Way are just moving away from each other (There are exceptions as well, but that is a different issue). Since then there were numerous cosmological models that have been proposed in order to explain this phenomenon of "Expansion Universal".

[And even our A1 made one. Alas, that was just a "trial and error" and one day he was so depressed and said that it was the "Greatest blunder of my life" when I presumed that universe is static that it is neither expanding nor contracting, provided with the fact that Izz had already told me that due to gravity of matter, universe could not be static.]

Now, let's imagine that you are back to your inky void and you discover that the very floor on which you are standing still has an intrinsic tendency of expansion, even with an accelerating pace. Or, in a simpler way, let's say that you are back to your inky void but now you observe that the very surface on which you are standing still or the floor underneath your feet is continuously expanding. What effect will it cause to your idea of mechanics in general and idea of motion in particular? Well, of course, you don't need me to tell you this, it will shatter whole of your existing understanding of motion in the universe. As if the very surface that you could use as a medium for your locomotion and the very surface that you could use as your coordinate system or reference frame, itself is expanding like a flat two-dimensional (2D) rubber sheet that expands uniformly in all directions and your present scenario in this inky void might look something like this.

This is what exactly happening out there in outer space and the only difference in these two voids viz your inky void and outer space of our real void, that is expanding in every direction is that the expansion in your inky void is 2 dimensional (2D), whereas the expansion of the universe in outer space is 3 dimensional (3D) or even 4D.

Now, let's say you made it possible, by developing a technology to travel from outer space to anywhere you want. And when you come back from outer space on Earth, you have an understanding, how space is expanding there or say you got the knowledge, how universe is synthesizing space in itself and now you are able to do the same here and what you do, you make the floor itself expandable as per your own wish or you got the technology how to create space within the space and it makes you "SPACER" because you are a super power now.

[I would like to emphasize the point that if you really take this idea of "Spacer" at its face value, it is for sure that you would be able to write a thesis on it or you can make a Sci-Fi movie or even you can write a science fiction book on this too. I tell you how.]

Suppose, you want to run faster than light, you don't need any vehicle for this anymore. *[Perhaps even it was this very vehicle that was your fundamental limit since the very first point of realization of yours that you need something, in order to move in the universe.]* And now what all you do is you just add some space within existing space and for the normal world, you would be far away, even you don't need an idea of motion now.

Good science fiction it might seem but it is just a metaphor in order to realize that the area of our investigation here in this book is far from being simple as its appearance offers us.

[You know what A1, you are a genius but it doesn't mean that whatever you say has to be correct all the time. For instance one day you said that

"On its basic level, nature is simple."
And even what "Occam's Razor" says

"........Simplest is the best ...."

But in contrast to this, I do realize that
"Nature has nothing to do with simplicity".
And the strongest support for this observation about the nature is "Chaos: An Emerging Science of Unpredictable".]

There is one more way to look at it. Say you are standing on the floor of this inky void and you observe that like a chess board, the floor you are standing upon right now is made up of two sorts of patches, "White Patch" and "Black Patch", like this.

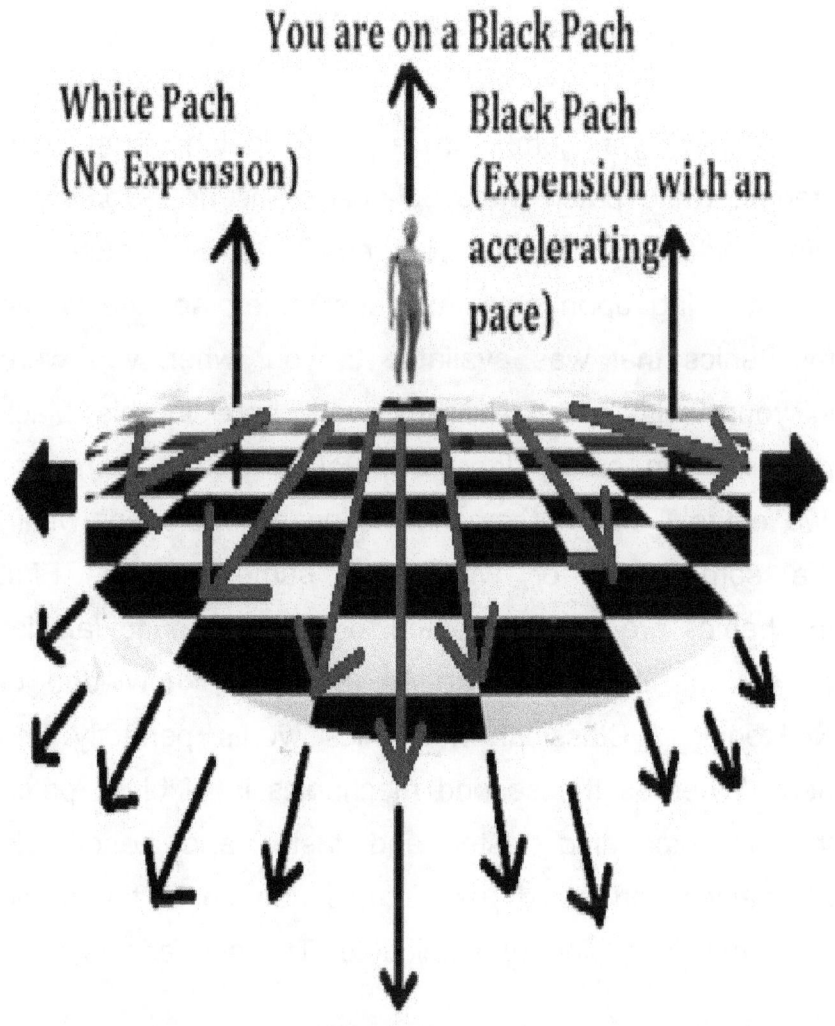

You are on a Black Pach

White Pach
(No Expension)

Black Pach
(Expension with an accelerating pace)

You now observe that in comparison to white patch, black patch expands faster (Let's say much faster with an accelerating pace.) and as this black patch is everywhere in the floor, so, you think that basically it is this black patch that is the reason for this expansion of your floor in all direction, exactly similar to a rubber sheet that is two dimensional and stretching in all the directions, as we saw earlier.

Now, you jump on a black patch, from your starting white patch that you observe is not expanding. What difference do you feel now? As the patch you are standing upon now, expands faster, so, the whole mechanics that was available to you, when you were in your white patch, doesn't work here in this black patch of the same floor. So, you see that there are two entirely different mechanics under operation, within the same floor or within the same domain. First mechanics are of white patch, one that is quite familiar to you, as it doesn't expand. So, your knowledge of Newtonian / Classical mechanics works perfectly fine here. Whereas the second mechanics is of black patch that is expanding faster and faster and hence old mechanics you could use, in this domain of the same inky void, is no longer applicable. The scenario is that

There are two sets of Laws of nature (or say Laws of Physics), one for the space that is not expanding, like your white patch in this inky void or space within our Milky Way galaxy or the space all around us here on earth surface, in your real void/world and second set of laws of nature for the space that is expanding faster and faster, with an accelerating pace, like your black patch here in the same inky void and the space between the galaxies in our real void/world that is this universe.

And as a matter of fact, we do have or we think we do have a little knowledge about the first set of laws of nature, what we call "Mechanics". But we know almost nothing about the second set of laws of nature.

Well, the idea is perfectly analogous to the current state of the expansion of the universe (of course, in a limit sense only), when the astronomical observations affirms that the outer space (space beyond galactic level or the space that is all around our Local Group galactic cluster) is expanding with an ever increasing pace (Quite similar to your black patch in the inky void). Whereas the inner space (space within the galaxy or even Earth) doesn't expand (that is similar to your white patch in the same inky void).

Where We Are Missing the Point? Well, because we have seen that the expansion of the universe is a 3 dimensional (3D) or even a 4 dimensional (4D) phenomenon, so, we really need to extend our analogy of "Expending Floor of An Inky Void" into the same 3D or 4D as well because our previous idea of a floor like an expanding rubber sheet, itself is limited to just 2 dimensions (2D).

So, imagine you are back in your inky void again but now something is different. You are in the same inky void but now you are experiencing a phenomenon, which as per your life long experience Is something "Supernatural" because when you are here on the floor of this inky void, you observe that this inky void is expanding in all the 3 dimensions, all around you, quite analogues to a situation where you are inside a large balloon and the balloon is expanding, as it is getting more and more gas from outside.

[Well, just to make things clearer, I need to emphasize that this "Expanding Balloon Analogy" as it is used throughout the globe for the explanation of "Expansion Universal" is far from the truth and it has perfectly no resemblance for the process that is naturally happening out there in space beyond Milky Way. But we are in a optionless situation because there is absolutely no analogy or examples we have, in order to describe or even compare the expansion of this universe.]

So, when you are in this "Expanding Inky Void", your true existence here is quite similar to a situation, when you are present in the "Intergalactic Space" where space is expanding all around you in all the possible physical dimensions with an ever accelerating pace. Something like this.

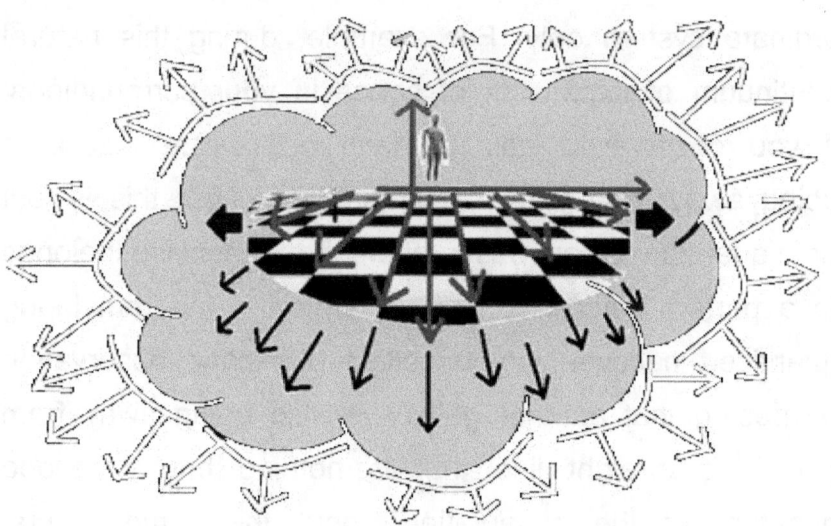

And I guess that it is the perfect time when I need to tell you why I am putting so much weight on the idea and science of "Observer". See, you are witnessing a continuum of expansion of space here, when you are in your "Expending Inky Void" or there, when you are in the "Intergalactic Space" in your real void. But it is the limit of human senses itself that you can never experience or observer any natural continuum whatsoever because like human vision for instance, whenever it observes a natural event, it always tends to quantize it, so that it can receive it and later on can extract a sense out of it. So, here as well, even if space around you is expanding like a natural continuum, still your observation about the expanding space gives you a sense that there is an expansion of space in reference to a certain co-ordinate system only. For example, during this natural continuum of expansion of space in your surroundings, if you observer a galaxy then you realize that it is going away from you in a straight line. But this is just one quantum of an observation that otherwise belongs to a perfect natural continuum, which is far from being quantized because if you collect a second observation by seeing that another galaxy is also going away from you in a straight line that is nothing but a second quantum of the observation about the same natural continuum and similarly by collecting few more quanta of similar observations, finally, you mix them up, quantum by quantum and conclude that space is expanding in every dimension, relative to you.

Just in case, if you would like to visit "Intergalactic Space", in order to see what this 3D (or 4D) expansion of outer space looks like, your existence there might appear like this.

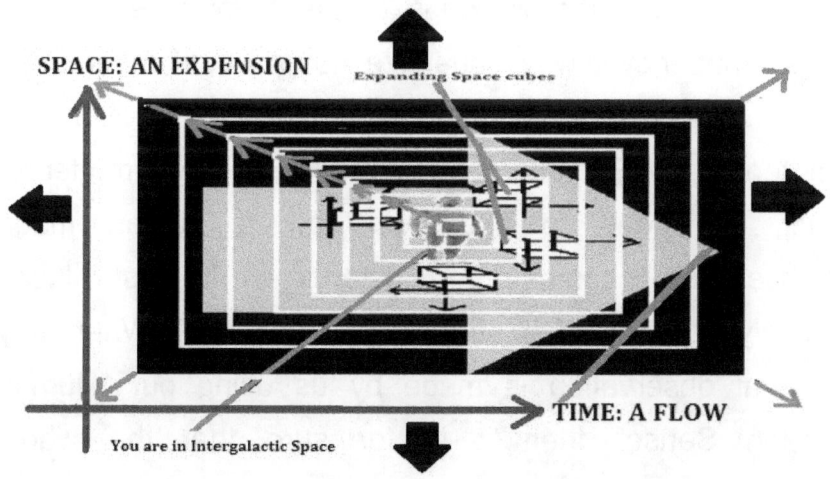

Well yes, it might seem confusing but unfortunately there is no way I can show you, what exactly is happening out there in intergalactic space on a piece of paper or even by using a 3 dimensional imaging but perhaps this is one of the reasons, why we invented this "Inky Void" at first place besides I really think that you must come out of the space that exists in your surroundings because on the name of "Space" whatever it is, it is just killing every truth for you and creating a platform for the landing of "The Matrix" that in turn creates a "Field" all around you, so that the "Power of Appearance" could be operational (Field like "Higgs Field"? Who knows?). So, it is only now, when you are out of natural space or the space that is given to you and enter into this inky void and begins your search for the ultimate truth.

Because the workings of the human senses are such that they always quantize natural stuff in order to make a sense out of it. For example light is quantized in terms of "Photons" as its particles. As a matter of fact, we need to find out, how a human made "Observation" comes into existence at its first place? Simply put, how we see things in nature? When any natural observation is made by us using our "Human Vision Sense" then it is for sure that the whole observation might fall on the retina (or human eye) all at once but it reaches to the back of our brain, quantum by quantum and hence somewhere in the journey, it gets quantized and simultaneously loses its intrinsic truth, which it supposed to carry to our brain.

So, when you are in an inky void and the void or this darkness that is present all around you is continuously expanding then this whole natural continuum remains beyond your understandings but if you are in the "Intergalactic Space" and the same void and darkness present in your surroundings is expanding then in reference to the existence of other galaxies in terms of a fixed "Celestial Co-ordinate System", this whole natural phenomenon makes a perfect sense to you.

Moreover, this intrinsic truth of the nature (or the absolute tendency) of human senses that makes every individual observation in this creation biased and a "Limit Concept" and creates a perfect layer that covers every scientific experiment or astronomical observation and even every single scientific equipment that can be used in our measurement processes and makes them all biased as well as a limit concept and this is how the truth always remains one layer ahead of "Power of Appearance". But this layer is like an "Absolute Interface" or an "Absolute Layer" (Something "Unbreakable"? May be.), which in turn creates a perfect isolation of "The Truth" from "The Appearance". Do you know who said this?

"Whatever is apparent is not true; whatever is true is not apparent."

[Although he didn't tell but now we know why and how it is happening to us. Don't we?]

So to conclude, if it is a matter of the power inherent in the idea of "Observer" then it is for sure that it is this power that makes every single observation and scientific experiment biased and limited and hence keeps it far from its underlying truth.

Even there is a probability that the observation that this universe is expanding (That in turn is an astronomical observation whose components as we saw are biased and limited), might just be a trick of the same "Power of Appearance" and hence there is a probability that it is our very act of making an "Astronomical Observation" through our telescopes that gives us an appearance of an expanding universe but the truth is still far more mysterious than the simplicity of the conclusion of this astronomical observation offers us. And hence it is quite probable that it is not the universe that is expanding but there is something else, who is the operator of this "Power of Appearance". And hence expansion of the universe, as it is believed today by the scientific community around the world is mere a "Mirage Effect" or a brilliant "Optical Phenomenon".

In order to support this highly unbelievable outcome, I tell you what. Even in 1929, when the discovery that "........the universe is expanding......", actually happened in the field of science, discoverer himself said that

"..........if redshifts are not primarily due to velocity shifts the velocity-distance relation is linear, the distribution of the nebula *(Galaxies of our time)* is uniform, there is no evidence of expansion, no trace of curvature, no restriction of the time scale and we find ourselves in the presence of one of the principles of nature that is still unknown to us today, whereas, if the redshifts are velocity shifts which measure the rate of expansion, the expanding models are definitely inconsistent with the observations that have been made, expanding models are a forced interpretation of the observational result......"

[Edwin Hubble, Ap.J. 84, 517, 1936]

He further reported that

" [If the redshifts are a Doppler shifts] the observation as they stand lead to the anomaly of a closed universe, curiously small and dense, and it may be added suspiciously young. On the other hand, if the redshifts are not the Doppler effects, these anomalies disappear and the region observed appears as a small, homogeneous, but insignificant portion of a universe extended indefinitely both in space and time.

[Edwin Hubble, Monthly Notices of the Royal Astronomical Society, 97, 506, 1937]

Besides, if you think a bit deep about your true existence in "Intergalactic Space" (or even in our "Expanding Inky Void" if you like) then the situation might appear like this.

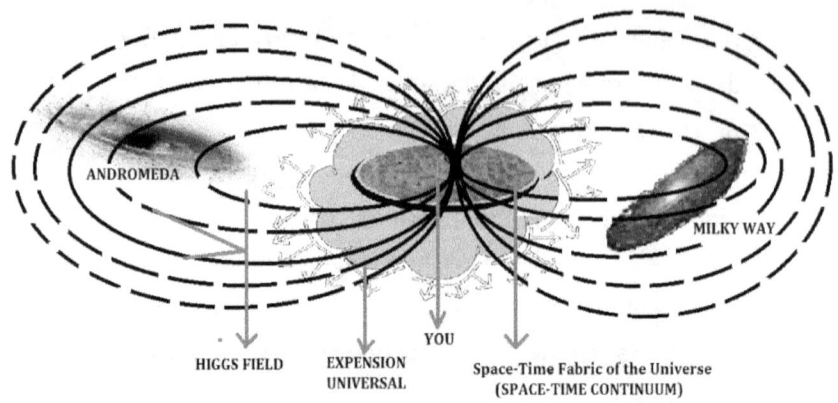

As you can see here that if you are in Intergalactic Space, there are three powers of the universe that are under operation, all around you viz Space-Time Continuum, Expansion Universal and Higgs Field. Moreover, if we devise an equation for the derivation of a natural truth here then it might look like this.

{Higgs Field} + {Space-Time Continuum} + {Expansion Universal} = {TRUTH}

And it is this equation that might give you a natural outcome that all these three different powers of this cosmos are in fact different manifestations of a single reality and it is this "Reality" that might include "Dark Matter" and "Dark Energy" too.

So, the simplicity of the conclusion that "Universe is expanding" based on the astronomical observations, since 1929 till date, might be elusive.

Like the process through which an astronomical observation is made, itself originates the appearance of an expanding universe, perhaps within the telescope, even before reaching it into our eyes [Or this transformation of "The Truth" (that a natural observation carries with it) into "An Appearance" happens within the human eye or even inside the human brain itself.]. Well, Of course, this whole point seems highly unlikely but if it contains even a single iota of truth, it simply creates another probability that every single astronomical observation ever made is corrupt due to this "Power of Appearance" and "Observational Astronomy" itself is unreliable and hence whole body of "Modern Astronomy" is compromised, no matter how much advanced technology you use.

[Or is it technological advancement itself that is the true source of this chaos? Because as we have seen in the words of Ed, at the time of the discovery of "Expansion Universal" itself, there were two distinct probabilities about the nature of this universe but perhaps it was this advancement of technology itself due to which one probability was killed outright and the second probability became science of today.]

[0]

n order to make our present idea of Absolute Motion closer to modern science, let's talk about something that is a subject matter of debate even today and before the advancement of science and technology, it used to be an area of intense discussion among scientists and philosophers and today it is called as "Absolute Time" and "Absolute Space".

Originally introduced by Izz in Philosophiæ Naturalis Principia Mathematica or what is known popularly as 'Principia' in 1687, the concepts of absolute time and space originally provided a framework for the development of Newtonian mechanics, as you can see right here.

surface of the earth, where the accelerative gravity, or force productive of gravity, in all bodies is the same, the motive gravity or the weight is as the body: but if we should ascend to higher regions, where the accelerative gravity is less, the weight would be equally diminished, and would always be as the product of the body, by the accelerative gravity. So in those regions, where the accelerative gravity is diminished into one half, the weight of a body two or three times less, will be four or six times less.

I likewise call attractions and impulses, in the same sense, accelerative, and motive; and use the words attraction, impulse or propensity of any sort towards a centre, promiscuously, and indifferently, one for another; considering those forces not physically, but mathematically: wherefore, the reader is not to imagine, that by those words, I anywhere take upon me to define the kind, or the manner of any action, the causes or the physical reason thereof, or that I attribute forces, in a true and physical sense, to certain centres (which are only mathematical points); when at any time I happen to speak of centres as attracting, or as endued with attractive powers.

SCHOLIUM.

Hitherto I have laid down the definitions of such words as are less known, and explained the sense in which I would have them to be understood in the following discourse. I do not define time, space, place and motion, as being well known to all. Only I must observe, that the vulgar conceive those quantities under no other notions but from the relation they bear to sensible objects. And thence arise certain prejudices, for the removing of which, it will be convenient to distinguish them into absolute and relative, true and apparent, mathematical and common.

I. Absolute, true, and mathematical time, of itself, and from its own nature flows equably without regard to anything external, and by another name is called duration: relative, apparent, and common time, is some sensible and external (whether accurate or unequable) measure of duration by the means of motion, which is commonly used instead of true time; such as an hour, a day, a month, a year.

II. Absolute space, in its own nature, without regard to anything external, remains always similar and immovable. Relative space is some movable dimension or measure of the absolute spaces; which our senses determine by its position to bodies; and which is vulgarly taken for immovable space; such is the dimension of a subterraneous, an æreal, or celestial space, determined by its position in respect of the earth. Absolute and relative space, are the same in figure and magnitude; but they do not remain always numerically the same. For if the earth, for instance, moves, a space of our air, which relatively and in respect of the earth remains always the same, will at one time be one part of the absolute space into which

[Page-77, "Mathematical Principles of Natural Philosophy" by Sir Isaac Newton-Published by Daniel Adee, 45 Liberty Street, New York]

According to him, absolute time and space respectively are independent aspects of objective reality and it was Izz who told us that

Absolute, true and mathematical time, of itself, and from its own nature flows equably without regard to anything external, and by another name is called duration. &

Relative, apparent and common time is some sensible and external (whether accurate or unequable) measure of duration by the means of motion, which is commonly used instead of true time.

According to Izz, absolute time exists independently of any perceiver and flows at a constant pace, throughout the universe as if,

There is an "Absolute Clock", somewhere in this universe.

Unlike relative time, he said that absolute time is just not the matter of human senses and only mathematics can help us to understand it. According to Izz, we are only capable of perceiving relative time that is a measurement of perceivable objects in motion. From these movements, we infer the passage of time, what Sir Stephan Hawking in his best seller "The Brief History of Time" called "Arrow of Time". Izz also told us that

Absolute space, in its own nature, without regard to anything external, remains always similar and immovable. &

Relative space is some movable dimension or measure of the absolute spaces; which our senses determine by its position to bodies and which is vulgarly taken for immovable space.

Now, our point of interest as per Izz,

Absolute motion is the translation of a body from one absolute place into another. &

Relative motion is the translation from one relative place into other.

What it suggests to us is that "Absolute Space" and "Absolute Time" don't depend upon physical events, but are foundations of cosmic reality, within which physical phenomena or events occur. Hence, every object such as you, me, this room, this book, etc has an "Absolute State of Motion" (an Absolute Motion?), relative to absolute space, so that an object must be either in a state of "Absolute Rest", or moving at some "Absolute Speed". But we have seen already that "At rest" is just a natural impossibility.

This Newtonian theme might be helpful in order to understand how a physical existence moves from one point to another in this universe and what we have discussed so far about Absolute Motion. And the final conclusion of this whole subject is that "Within the existing framework of modern science, there are two frames in nature. First is "Relative Frame", in which everything is relative, such as space is relative, time is relative and motion is relative. And the second is "Absolute Frame", in which everything is absolute, such as space is absolute, time is absolute and even motion is absolute. Let's draw a diagram of this truth cosmic.

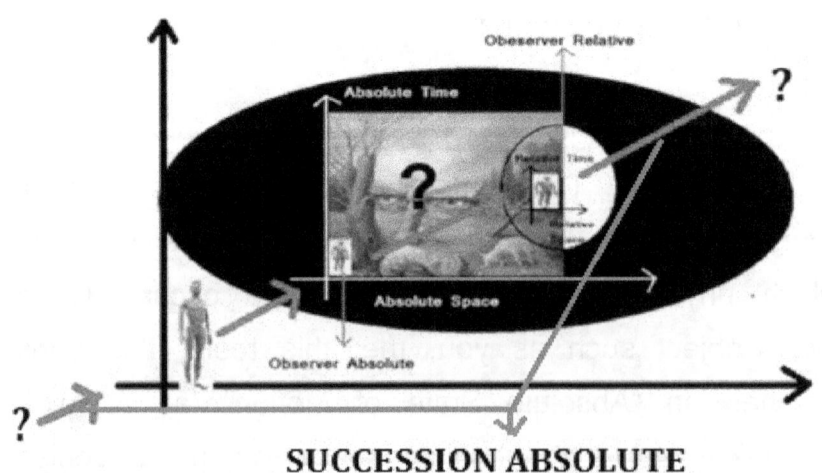

SUCCESSION ABSOLUTE

[There are pairs like "Absolute Vs Relative", that is Newtonian, "Proper Vs Improper" that is Einsteinian and "Actual Vs Potential" that is Aristotelian and important here is that when they talked about these pairs, they were thinking about the same thing and the thing was "nature" or this "universe".]

Now let's talk about first frame that is "Relative frame", a frame of reference or a coordinate system or a "Relative World", in which every individual existence is relative. And the second frame is "Absolute frame", a frame of reference or a coordinate system or an "Absolute World", in which every single existence is "Absolute". For instance, let's say that our home galaxy "Milky Way" is a relative or proper or potential universe and the universe that includes observable universe within it is an absolute or improper or actual universe. In this way it is obvious that there is an interaction between these two universes namely,

"Absolute Universe Vs Relative Universe"

Or

"Proper Universe Vs Improper universe"

Or

"Actual Universe Vs Potential Universe".

BLUE BOOK

[P]

E ither it is Higgs boson or our quark (or it is a lepton or boson as Standard Model says), what seems certain is that universe is made up of particles. So, the fundamental building block of creation is particle and we need a true realization of the concept of a "Particle" and the best way to do this is to frame a question,

What is particle?

Moreover, we have already seen the power inherent in the idea of "Particle", as our first puzzle is "Illusion of Solidity", second is the curiosity that "What the world is made up of?" and an appearance that the fundamental domain of inertness of a physical in this universe, what we are saying as 'Inertia' is a particle itself, third, our realization that the ultimate source of this creation that looks so solid and perfectly obedient for the laws of nature here is basically coming from an ocean of chaos or we should say a quantum mechanical chaos, as we saw earlier in the "Green Book" and finally our conundrums of

Particle within the Particle?
Or,
Box within the Box?
Or,
Appearance within the Appearance?
Or,
Chaos within the Chaos?

Now, the question is

"Does this 'Dawn of 21st Century Human Civilization' have a true Definition of a particle?"

[And just in case if you don't know. In a sort of "Rat Run" of making "Big Science", India itself is planning to construct her own "Particle Accelerator" that is a billion of dollars project and if we don't have, even a true definition of a "Particle". This whole future Indian investment in form of a "Particle Accelerator", instead of helping millions of people, who are dying from poverty and chill, will be nothing but a perfect crime. Well of course, this is the truth for CERN and its LHC (Large Hadron Collider) as well, as we have

seen in "Green Book". But we have nothing to do with them. Or do we?]

Well, I doubt this whole stuff and this is the reason, I would like to present you a new idea about this concept of "Particle" and as a matter of fact, I am redefining a "Particle". So now, we redefine a particle.

"A particle is a quantum of the field, in an instant."

In more elaborated sense, particle is,

"A particle is a quantum of the field (to which it belongs), in an instant (in which it is measured or detected)."

Or, even more precisely, a particle is,

"A particle is a quantum of the field, to which it belongs, in an instant, in which it acquires its relativity".

Modern science in general and modern physics (Particle Physics) in particular defines a particle as

"A body whose spatial extent and internal motion and structure [if any], are irrelevant in a

specific problem."

[http://www.thefreedictionary.com/particle]

Or,

An elementary particle. A very small piece or part; a tiny portion or speck.

Or,

A very small or the smallest possible amount, trace, or degree.

An elementary particle, a subatomic particle.

[dictionary.reference.com/browse/particle]

If you can see the efficiency of this new proposal, we are saying "n-DOP" (new-Definition of Particle) for a true realization of the idea of a particle lies in the fact that it can be applied on all the three manifestations of this nature namely quantum, classical and cosmic.

And I am going to show you how. Now, what if we combine "Feynman Diagram" (A pictorial representation of the mathematical expressions describing the behavior of subatomic particles) with the idea of an "Arrow of Time" and the concept of "Cosmic Microwave Background" (CMB)? This might look like this.

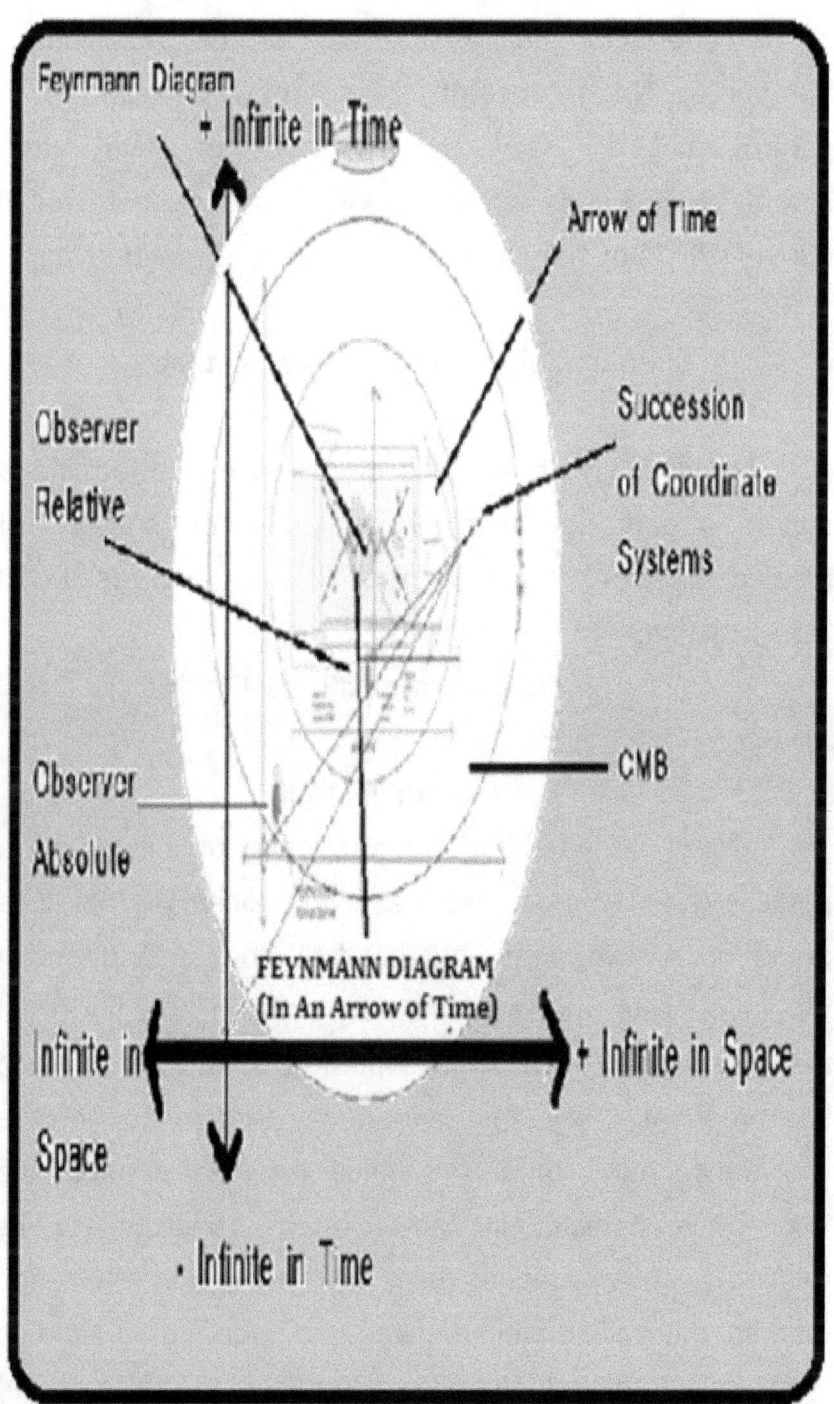

It perfectly justifies the use of the inclusion of the phrase "In an Instant", in our new definition of a particle (n-DOP). Now, for the sake of clarity (that automatically brings simplicity with it) we need to recall our n-DOP. So, here we go. A particle is defined as,

"A quantum of the field, in an instant."

Now, if you break this definition by performing a minor surgery (a physical analogue of surgery of medical sciences of course) on it, the end result will look like this.

"A Quantum of the Field

+

In an Instant."

Now, the testing of our new definition of a particle (n-DOP) becomes easy. We observe that it is made up of two parts, namely first and second and if we investigate the subject deeper then we find that there is nothing new, in this portion of the first part of this proposal. Okay, that is perfectly true (although in a limit sense only), modern science when talks about the idea of a particle, in terms of its relativity (or its relation) with an idea of field, we do have a realization that an interrelationship between the idea of a field in terms of a particle (its own particle to be precise) exists, although not in so obvious sense like "A field based or a field dependent definition of a particle" or "A particle dependent definition of the field" but there is a scientific realization that goes like this, "Field and its associated Particle", somewhere in the standard text on the subject. Now, the main point comes that is definitely the soul of this proposal, we are calling as "n-DOP". What about the remaining second part? Namely,

"In an Instant."

This is new. Isn't it? And the soul effort to make our discovery what we termed as "n-DOP" scientific is definitely visible in our attempt to combine, three concepts of modern science, viz "Feynman Diagram", "Arrow of Time" and "Cosmic Microwave Background" (CMB).

So, discovering a solution for this paradox, what they call the unification of Quantum Mechanics with General Relativity (so called the "Quantum Field Theory"), "A Holy Grill of Theoretical Physics" is a perfect human endeavour provided we are ready to re-examine our existing concept of particle itself and its relation with the field. And this whole unification might look something like this.

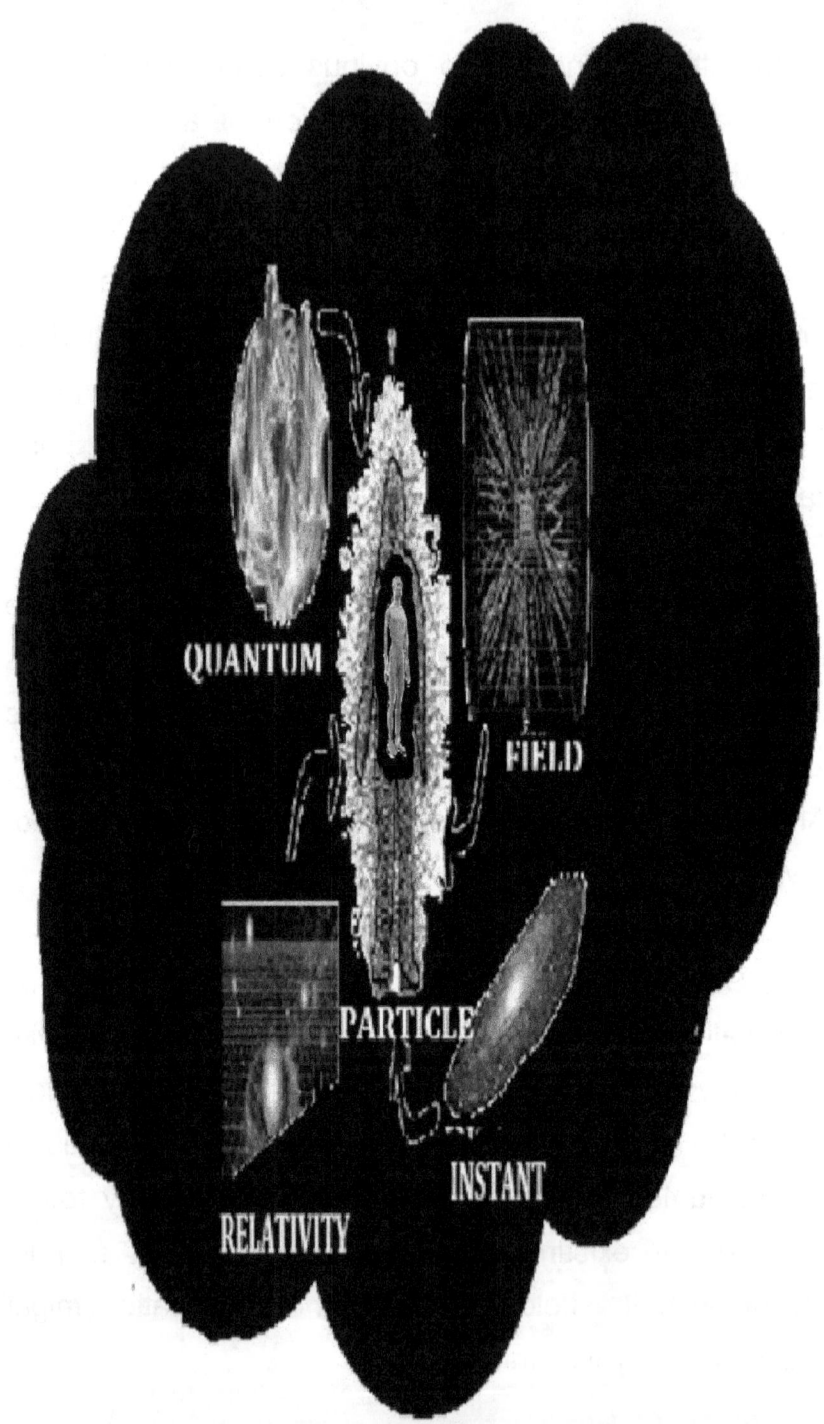

Because, if power of all these titans viz "Quantum", "Field", "Instant" and "Relativity" gets unified, they originate a power of first order that is a "Particle", a power that creates everything in the cosmos, like a fundamental building block of this entire creation or a brick of this "Grand Design".

I would like to present you an instantaneous application of this new definition of particle (What we are saying "n-DOP") to define our home in this apparently infinite universe, we call Milky Way Galaxy. As expected, this new definition (n-DOP) must give a clear picture to which it gets applied upon, which we have seen can't be achieved by our old definition of particle given by modern science. So, we use our new tool, we say 'n-DOP' in order to define Milky Way as a cosmic existence, with a general assumption that if it is a matter of a reference frame, this object that otherwise is a perfect galaxy in this universe, in some particular context of scale can be presumed as a particle or even as a matter particle and then we come up with this new idea. Milky Way, our home galaxy, is defined as,

"A quantum of a field called as space, in an instant, called as age of the Universe."

In a more elaborated way, a particle called as a galaxy and termed as Milky Way is defined as,

"A quantum of the field, called as 3 dimensional Space or a 4 dimensional Space-Time, in an instant, called as age of the universe, when it acquires its relativity, in terms of its cosmic existence that is called as galaxy with respect to this observable universe."

Moreover, as we have discussed in White Book, there is a discovery of "The Great Attractor", what is called as a gravitational anomaly, somewhere 150-250 million light years away from Milky Way, and we are dragging in toward this point. So, if we redraw this cosmic scenario, in a very simple diagram like this.

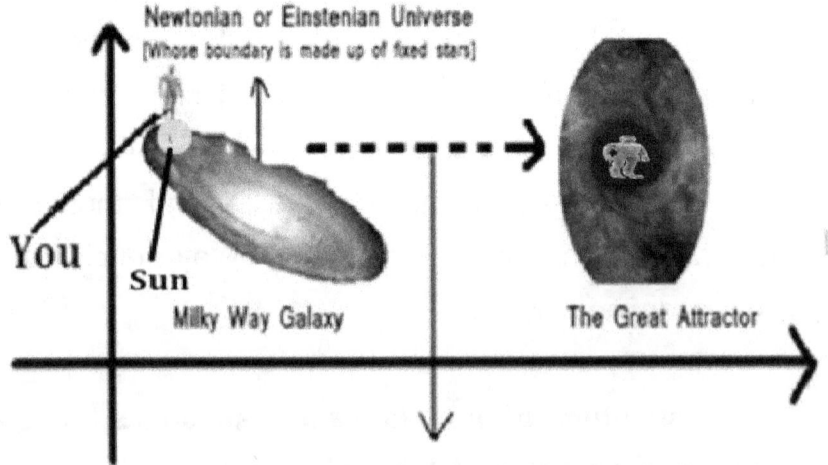

World Line of Milky Way in This Cosmic Phenonenon

Our starting idea of presuming Milky Way, as a particle in its field becomes obvious. Because now it's "World Line" itself is visible.

I must confess that we never really needed to study this whole subject because this is all about "Particle Physics". But I did, what I did, because I had to. You see, without a true idea of a "Particle" or a "Cosmic Particle" or a "Creation Particle", it was not really possible for us to continue our journey, in search of an ultimate truth.

Because you really don't need me to realize that if we don't have a true definition of a particle (DOP) or if we don't have a true realization of an idea of a particle, then it is for sure that this absolute "Cosmic-Quantum Conundrum" can never be resolved. And the best example I have for you, in order to prove this point is our existing knowledge or science, about the idea of a natural particle, what we call today as "Modern Particle Physics" because if they could do it, then the truth of this puzzle could be as obvious as our Sun itself.

But it is certain that since the ancient time till date, nobody ever was successful to resolve this fundamental problem of humanity.

DOOMSDAY

N ow, let's talk about "Doomsday" or simply the "End of This World". You see, construction of a house starts from the idea of "A Brick" and it is true for this whole creation as well. Moreover, because idea of creation starts from the idea of a particle, so, the idea of destruction must also depend on the same idea of a particle and a true realization of the idea of a particle, like "How a particle comes into existence in the universe?" with an equally important question "How a particle goes out of existence? That can be used as a key to understand the "Science of Doomsday". It appears that the phenomena of "Moment of Destruction" will follow the exact mode of "Moment of Creation" but exactly in the reverse direction. So, if we approach Doomsday in the universe then,

First our galaxy Milky Way will decay (This phenomenon is already going on the edge of Milky Way), followed by the disintegration of Solar System. After that Earth will disintegrate and then there will be decay of complex molecules like DNA and then there will be a decay of proton and then quark will decay. And whatever exists beyond the quark (if it does exist) will follow the same pattern of destruction and this process will go on towards the limits of infinitesimalness, till the end of the same absolute stuff from which the ultimate construction of this "Grand Design" begun.

[So, there is really no need to waste millions of dollars, just to observe the "Decay of Proton" as our Big Science is doing today, just because the time has not yet come.] Despite of the fact that what modern science has to say about it, it seems that the "Science of Moment of Destruction" that is our Doomsday is intimately linked with the "Science of Moment of Creation". Hence the true knowledge of half the subject will be enough to complete whole science on this subject. So, the mode by which this cosmic-quantum puzzle of "Particle within the Particle" or "Box within the Box" came into existence in this universe will be resolved by the same mode but just in a reverse direction. Moreover, there is another possibility,

"What if Milky Way itself is Higgs Boson?"

As it is the case with Spontaneous Symmetry Breaking or what we discussed about the transformation of Supersymmetry (SUSY) into Broken-Supersymmetry (B-SUSY) like a phase transition, when water gets converted into ice where the cause of the pattern formation or chaos itself always exists within the object. So, for an outside observer it looks arbitrary, such as spiral geometry of the Milky Way looks arbitrary to us.

[Q]

Imagine, you are standing outside of the observable universe and from outside you observe that this observable universe is quite similar to a block or a matter particle. Something like this.

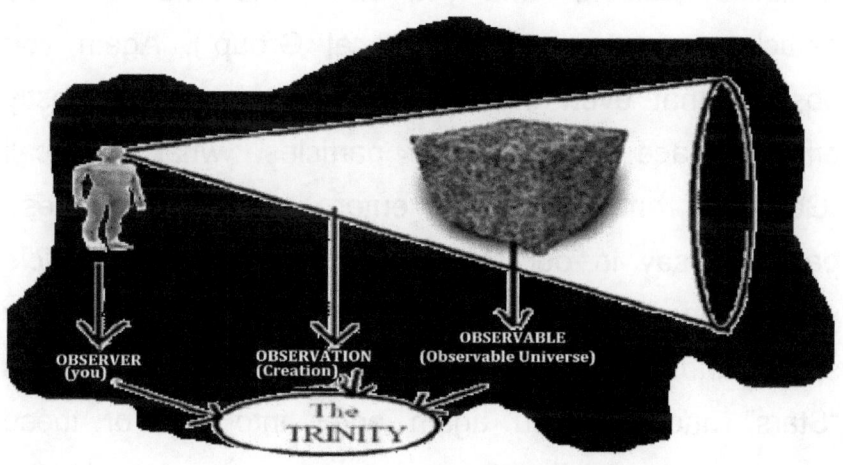

[Here it is obvious that there is a perfect trinity (The Trinity?). And what is even more important is that there is a trinity of "3 Os" namely "Observer", "Observable" and "Observation" and entire observable universe exists within this trinity. And the top most fundamental in this trinity is "Observer" that is you.]

And now you enter into this particle or observable universe and then you observe that this particle is made up of mostly empty space and smaller particles inside of it, what we called as "Galactic Superclusters" and you enter into one of these particles [Let's say in our "Virgo", Local Supercluster, because if you don't then nobody knows, where you will be in this infinite darkness.] and now you observe that this particle itself is made up of mostly empty space and smaller particles again, what we say "Galactic Clusters" and you enter into one of these particles now (say in our "Local Group"). Again you observe that even this particle is made up of mostly empty space and smaller particles, what we call "Galaxies" and now you enter into one of these particles (say in our "Milky Way"). Inside this particle you observe that it itself is made up of mostly empty space and smaller particles again, what is known as "Stars" and say you again enter into one of these particles (say in the Sun). When you are inside the Sun (or Solar System to be precise), you find out that it is also made up of almost empty space and of smaller particles, what we say "Planets" and now you enter into one of them (say in Earth). Now, from inside of the Earth, you observe that it itself is made up of mostly empty space and smaller particles again, what we say "Molecules" and you enter into one of these particles and find out that it is made up of mostly empty space and even smaller particles again, called as "Atoms" and now you enter into one of these particles and from inside of an atom you find that it is made up of mostly empty space and even smaller particle called as "Nucleus" and if you enter into a nucleus of an atom, you observe that even this is made up of mostly empty space and smaller particles known as Protons and Neutrons and now you enter into one of these particles and you find that it is again made up of mostly empty space and smaller particles called as "Quarks". And now........................... ad infinitum?

[Well, if you are thinking that perhaps I am repeating the same stuff, again and again, since the beginning of this book then I am afraid, you are absolutely right. But it is so because we need to focus and also because of the fact that the subject of our investigation, in this book is far from being simple, as they have told you.]

Now see the beauty. In this whole journey of yours, it is certain that your every single realization, whenever you made it, about a particle, was an illusion. Because every time when you entered in a particle, you observed that your previous conclusion that this is a particle was proved wrong. As it is just like an almost empty space. And this is a quite efficient approach, in order to show you the potential inherent in the idea of "The Matrix" and see the beauty of the matrix.

There exists not even a single boundary or limit that could isolate two particles from each other. So, whatever exists, it exists in a perfect continuum and the beauty is that still everything looks perfectly isolated from the others.

And it is for sure that one of the most fundamental powers of this matrix is "Surface" (or more precisely an "Interface"). So, this matrix of this creation or universe has an inherent "Power of Surface" and a "True Surface" exactly like "True at Rest" doesn't exists anywhere in the universe. And whatever exists, it exists in a perfect cosmic continuum, from the cosmic horizons towards the limits of observable universe up to the quantum limits towards quark. This idea is metaphorical in the sense that now we have a new hypothesis that says that if you leave conventional idea of "Dimension" in terms of 3 Dimensional (3D) space or 4 Dimensional space-time (4D), it appears that on its fundamental level,

"Universe is two dimensional (2D)."

First, that starts from human retina or optic chiasm or simply back of your brain, towards the limits of observable universe and second that starts from the same human retina or your brain, towards quantum world of quarks and gluons. And your existence as an observer is exactly at the midpoint of this "Cosmic-Quantum Dimensionality".

It implies that we must revise our existing understanding about the concept of dimension, as it appears that an observer dependent approach, towards the true understandings of the idea and science of dimension says that

The universe is basically two dimensional and our conventional three dimensions are nothing but sub-dimensions of these two fundamental dimensions.

Let's say that the first dimension of the two dimensional world or universe, starts from human retina or simply from human eye up to the limits of observable universe towards the cosmic horizon is "D" and the second dimension that starts from human eye up to the quark level towards the quantum horizon, what we have discussed about "Quantum Foam" or "Space-Time Foam" is "D^{-1}" (D Inverse). So, the situation appears something like this.

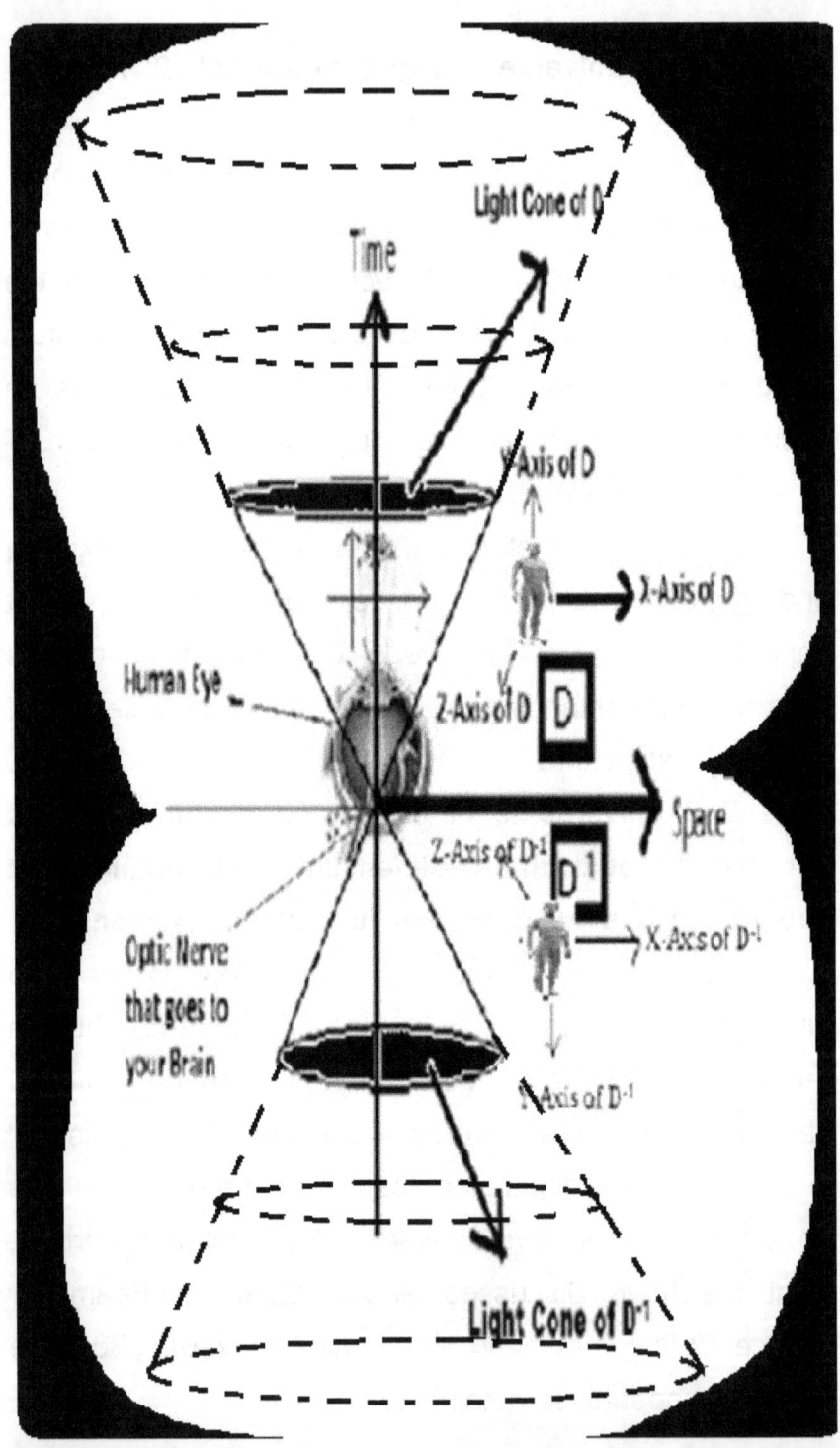

So, the universe is existing within an open interval of "+ Infinite" to "- Infinite" [or {+ ∞, - ∞}] and the existence of an observer (Like you) of this interval is exactly at the point on which these two infinities meet. And it might look like this.

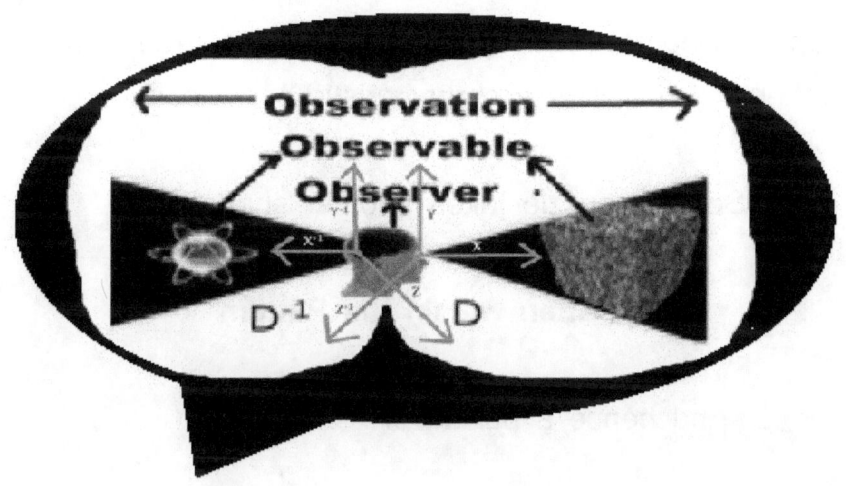

Or, it might appear like this.

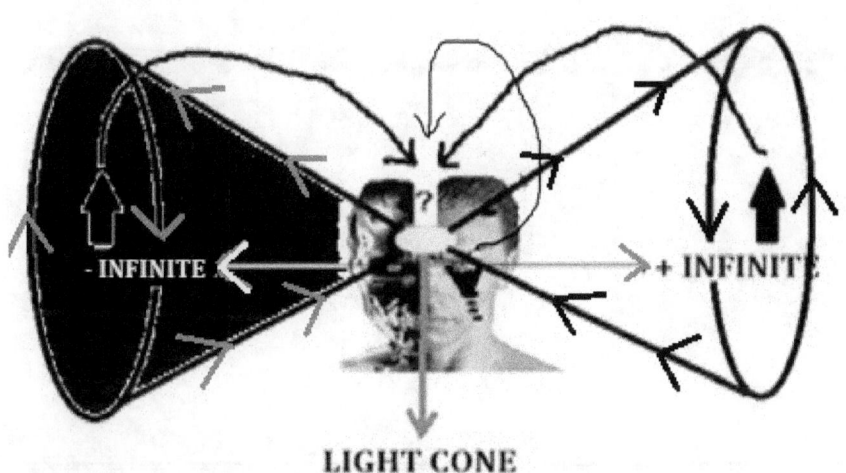

Moreover the Geometry of our galaxy milky way is so impressive in the sense that you take a phonograph and place a tennis ball at its centre. The geometry you get from this is quite similar to our galaxy's geometry that gives us one more clue that there is another puzzle within this puzzle because the "Spiral Geometry of Milky Way" is like a pattern but pattern formation is a characteristic of "Chaos".

So, this puzzle takes another shape of

"Pattern within the Pattern"

And hence a puzzle of

"Chaos within the Chaos".

[R]

I t will be quite difficult for us, if we directly jump over the subject because it is pure mathematics. But I assure you that the base trick is simple and once you get it in your mind, whole concept with its mathematical beauty will be yours. So, it would be quite logical, if we go with the fundamentals first. Okay? Let me tell you something about "Fractal".

A fractal is a natural phenomenon (Please make a note of that because it appears that by creating fractals in nature, such as "Lightning bolts", "Heartbeats", "DNA", etc universe is doing its own mathematical calculations). Or,

A fractal is a mathematical set that exhibits a repeating pattern that displays at every scale. If the replication is exactly the same at every scale, it is called a "Self-Similar Pattern".

And as we saw, it is this "Self-Similarity" that also exists in our cosmic-quantum conundrum of

"Particle within the Particle". Fractals can also be nearly the same at different levels. This latter pattern of Fractals also includes the idea of a detailed pattern that repeats itself. In general, fractals are objects that

1.   **If you zoom in and look at one part of it, it looks the same as the whole.**
2.   **They have interesting structures, no matter how closely you look at them**.

Having said that, first we start from the basics and let's say it's a mathematical object, popularly known as "Canter set". There is one more mathematical object called as "Sierpinski Carpet", which we will talk about later on but basically this canter set is a one dimensional version of sierpinski carpet itself that is a 2 dimensional stuff. Because in order to get a Canter set, what all you need is "A straight line" of any length you desire.

So, now let's say that you have a straight line in front of you and suppose it is 9 inches long. And what you do, you divide this line into 3 equal parts. And you remove the middle third of this straight line. Now, again you take the remaining straight lines and divide them into 3 equal parts and you remove the middle third in every one of them. Like this you just keep on repeating the first step on the remaining straight lines...........ad infinitum.

This whole operation in order to get a cantor set must go quite similar to this.

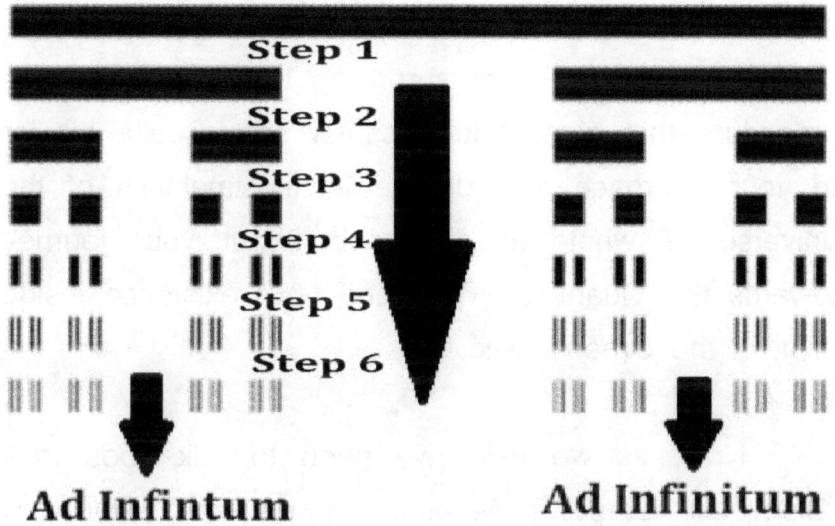

Step 1
Step 2
Step 3
Step 4
Step 5
Step 6

Ad Infintum   **Ad Infinitum**

[Well, the procedure through which fractals are made or created in nature demands your time and efforts for its proper understanding. So, just hold on, it is really interesting.]

Now, see the beauty, once you have removed infinitely many stages, each smaller section of this straight line, is exactly the same as the whole line. It is actually the procedure you followed that breaks the original line, into infinitely many pieces.

So you see, it is quite surprising because it is exactly similar to our original puzzle of "Particle within the Particle" or what we said "Box within the Box" analogy. Because within this canter set, a similar pattern is existing, like there is a straight line within the straight line within the straight line …………..ad infinitum or simply "Line within the Line".

And what is more surprising is the fact that the procedure that gives you a canter set is quite similar to your approach towards the infinitesimalness of the universe or what we discussed about your journey towards the Quantum World and your existence inside of it, in the Green Book.

Now, as we said, we need to talk about this "Sierpinski Carpet". Because in general nature all around us is 3D (or even 4D), so, we need to investigate pattern formation that is intrinsic to the idea of "Chaos" in 2 dimensions as well. Because your previous approach to get a canter set is limited to just one dimension. So, here we go with our Sierpinski Carpet. Mathematically "Sierpinski Carpet" is a perfect example of "Fractals" and as it was the case with canter set, we need to follow a new procedure that is quite similar to our previous procedure but now we are working with 2 dimensions (2D).

So, now we need to create a sierpinski carpet and again we follow a set of rules quite similar to the one that we used in order to get our cantor set. So, what you do, you take a square of any size, whatever you see fit and you divide this square into 9 smaller and equal squares and you remove the central square.

Again you follow the same procedure on the remaining 8 squares and you divide all these 8 squares into their 9 equal and smaller squares and then you remove the central square from all of them and again you divide the remaining 81 squares into their 9 equal parts and you keep on doing this................ad infinitum. And at last what you get is a "Sierpinski Carpet".

And the whole operation of yours must appear like this.

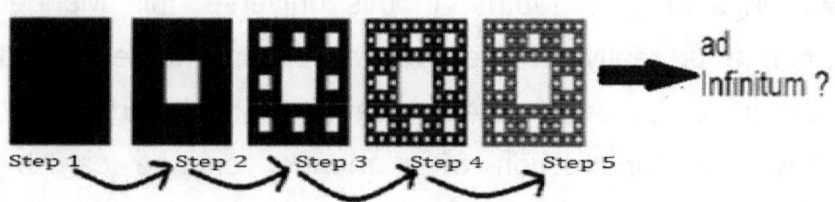

But the problem is that as we are living in a 3D space or a 4D Space-Time, so, the natural patterns are also created by nature in 3D. And this is the reason; we are talking about this mathematics of fractals and let's talk about the next concept.

In mathematics, it is called as the "Menger sponge", that is again an example of fractal geometry (And as a matter of fact, it is this mathematical object only, for which I wanted to talk about fractals, at its first place because there is something really beautiful in this mathematical object).

Menger sponge is basically a three-dimensional generalization of the cantor set and sierpinski carpet. What is really important in it for us is a fact that the menger sponge simultaneously exhibits an infinite surface area and zero volume.

I repeat "Infinite Area, Zero Volume". Because I do realize that in order to understand the fundamental workings of "The Matrix" in this universe, this Menger Sponge is really a nice tool, as nature herself could use it, in order to create patterns in the universe. Now, the construction of a menger sponge can be described as follows:

1   **Let's start with a cube.**

2   **Divide every face of the cube into 9**

squares and this will sub-divide the cube into 27 smaller cubes. Okay?

3   Now, remove the smaller cube in the middle of each face, and remove the smaller cube in the very centre of the larger cube, leaving 20 smaller cubes.

[And as you can observe, it perfectly resembles with a void that have a cube shape (make a note of that as this point is about to be used by us)].

4   As we did before, just repeat step 2 and 3 for each of the remaining smaller cubes, and continue to iterate ....ad infinitum.

And the whole procedure will appear something like this.

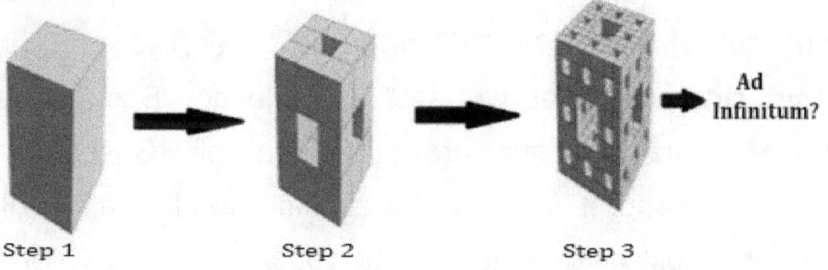

Step 1   Step 2   Step 3

[Look at this; the menger sponge itself is the limit of this process after an infinite number of iterations. Because there is no way that you could go ahead and perform the same procedure in 4D (Well, is it really impossible? Because what if we take an imaginary 4D Space-Time cube and continue this operation on it?).] And the final result of this whole operation is this.

Well, I am sure that you saw this structure somewhere, didn't you? It might not be exactly similar but quite similar. No? Think again. Yes, you got it. This structure what we said as "Menger Sponge" is quite similar to our previous idea of "Observable Universe" and it is really fascinating as I don't think that it is just a coincidence. Do you?

Moreover, you can see that all the ideas like "Fabric of the universe" in terms of quantum foam, our puzzle of "Particle within the Particle" and even the idea of "Illusion of Solidity" etc are present within this single geometry. Like this.

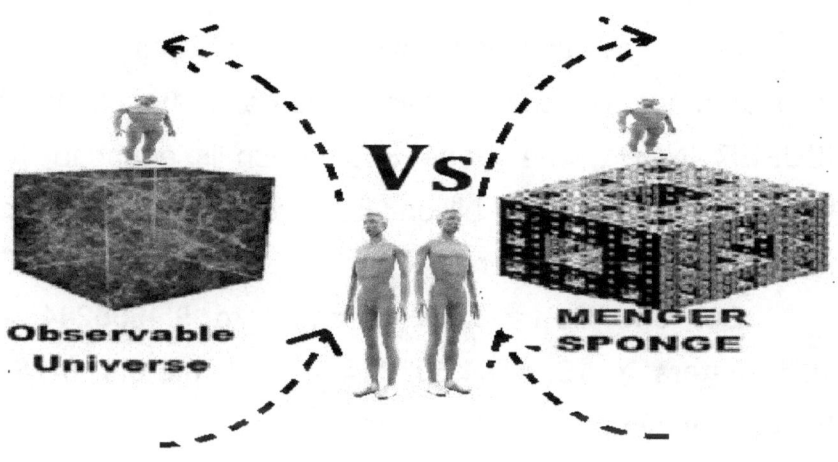

As we observed, the menger sponge is a closed set, since it is also bounded, it is compact. It is an uncountable set.

In 1926, it was Menger who showed, in his original construction, that the sponge is a universal curve, in that every curve belongs to "Homeomorphism" (Self Similarity?). To a subset of the menger sponge, where a curve means any compact metric space covering dimension one, this includes objects like trees with an arbitrary countable number of edges, vertices and closed loops, connected in arbitrary ways. In a similar way, the sierpinski carpet is a universal curve for all curves that can be drawn on the two-dimensional plane. And now comes our real point of interest.

The Menger sponge has an infinite surface area, yet zero volume.

Let's say that we invent a new unit for the measurement of cosmic distances and we term it "SUPER INCH", where one Super Inch is equal to 10 billion light years [1 billion light years is equal to 9.4605284×10^{24} meters].

So, one Super Inch is equal to 9.4605284×10^{24} meters $\times$ 10 $= 94.605284 \times 10^{24}$ meters of SI System or simply just 10^{26} meters.

And you take this cube, what we called "Observable Universe" and presume that it is a cube of 9 × 9 Super Inches [Why? Because if you recall, in "White Book", we have already seen that the recent measurements made by cosmic surveys predict that our observable universe (and here our cube) is approximately 90 billion light years across or if you presume that the observable universe is a cosmic sphere and you are exactly at the centre point of this sphere, its radius would be approximately 45.7 billion light years. So, what all you need to do is that presume that the geometry of observable universe is a cube of 9 × 9 Super Inches, instead of a cosmic sphere of radius of 4.57 Super Inches.

And now, if you perform the same procedure on this cube, exactly what we did in order to get "Menger Sponge" (as we saw that this universe is made up of mostly empty space in terms of "Space within the

Space" or "Darkness within the Darkness"), the end result will be a new menger sponge for sure, with the miraculous power of having infinite surface area yet zero volume. And this is wonderful. Keep going, because now, what you do is fill this Menger Sponge that is this cube of observable universe here, with some special kind of liquid or say a "Super Liquid". Even better if you just fill it with a "Quantum Field", exactly similar to the field, we have discussed in terms of "Higgs Field" and suppose now you enter into this cube. That might appear like this.

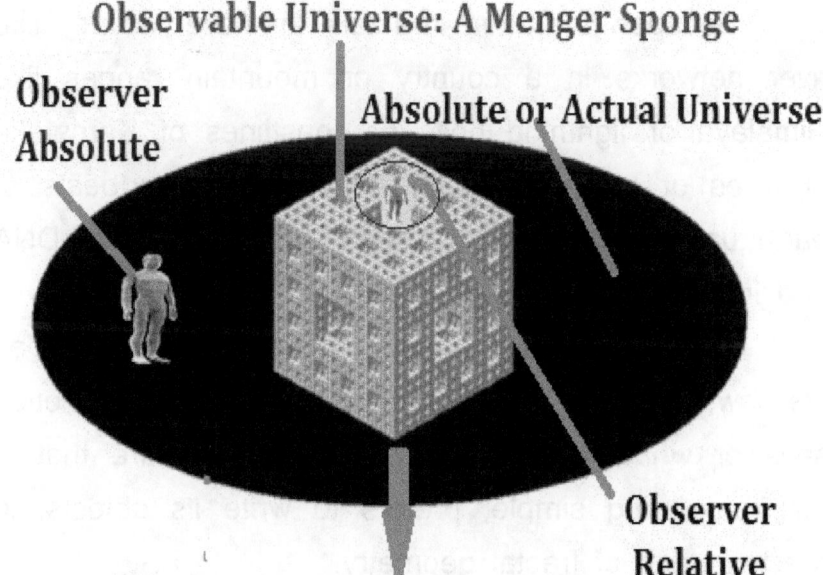

Observable Universe: A Menger Sponge

Observer Absolute

Absolute or Actual Universe

Observer Relative

A Universe with an infinite surface area, yet zero volume.

A Singular Point? Or, Universe: A Singularity?

What will happen? Certainly, because there is flow of Higgs field within this Menger Sponge that is made up of observable universe, you will see an endless series of patterns all around you, as they exist in the idea of Chaos or science of chaos, known today as "Chaology". An endless series of structures and there is a probability that there would be structures like spiral geometry of Milky Way and hence you will observe "Structure within the Structure" or even "Geometry within the Geometry", exactly analogous to our previous idea of "Self Similarity", when we were talking about "Fractals".

Because in nature fractals are everywhere, like river networks in a country or mountain ranges like Himalaya or lightning bolts or coastlines of Karnataka or trees or pineapple or heart rates or heartbeats or earthquakes or blood vessels or ocean waves or DNA and infinitely more.

And this is good enough to convince you that as we saw, despite of the fact that what Einstenian notion says or what "Occam's Razor" predict, nature that is far from being simple, prefers to write its objects, in the language of fractal geometry.

And I tell you what, we really need to go back in time to visit Euclid (300 BC), when he was preparing "Axioms of Geometry" that is exactly what we learn in our schools today in terms of straight lines, triangles, spheres or rectangles, etc what we call "Modern Geometry" and we could suggest Euclid that because fractals are the most natural geometry in the universe, so, probably he should take fractals as the foundation of his geometry in his book "Elements", the book that gave birth to our "Modern Geometry" called as "Euclidean Geometry".

Now, we need to talk about a very important scientific concept that is going to be utilized in the next chapter and a true understanding of this scientific tool is what we really need in order to reach the final conclusion of our investigation in this book and it is called as "Space-Time Continuum" and for the sake of simplicity, we will follow the following definition of space time continuum.

"Space-Time Continuum is a collection of parametric specifications that attempt to define "is".

Specifications can be as single value and thus define a specific point or they can be continuous to define entire entity."

[http://cosmicshipmedia.net/spacetime/spacetime.html]

[S]

In order to realize what Absolute Motion could look like to its observer like you, as it can't be observed directly, it is important to consider some facts discovered by modern "Observational Astronomy" such as, the average speed of the Sun in Milky Way is about 230 km/s and the estimated velocity of galactic cluster that is our Local Group is 600 km/s and satellite measurement has determined that if you take Cosmic Microwave Background (CMB), as a background i.e. a cosmic background as a reference frame,

Earth is moving with a velocity of 390 km/s towards a constellation "Leo" in this background that is CMB.

So, the idea of Absolute motion could be realized if you consider all these independent and relative motions of Earth and that in turn get integrated inside you as an observer, just because you are not an isolated entity for these motions. And the resultant of this calculation will give you a face of Absolute Motion, so that you would be able to recognize it.

Now, we need to come back to our main track and let's recall our conclusions that everything or every physical is in motion. So, we don't need to worry for the source that causes the interaction between the physicals and this Higgs field in order to come to an existence, as a massive and inertial object in the creation.

Because everything moves all the time and everywhere, so it could interact with this all pervasive Higgs field, everywhere and all the time. And hence it maintains its physical existence as a massive thing like Earth or a mountain or even a subatomic particle like protons or a quark, everywhere and all the time.

Well, you really need to go back to your inky void now (you are already there anyway) and realize these two deductions of our reasoning that everything is in motion, what we called as "Absolute Motion" just because of its cosmic continuity and another one is this theory that when a physical object/particle/existence interacts with Higgs field, it is this interaction that makes the body massive or perfectly physical (and hence inertial). What all you need to do now in this void is that just replace this physical object by your own human body because your body is similarly as physical as any other thing. So, now the scenario is this.

You are in this inky void alone, nothing else is existing (except your body?) and you are in a continuum of motion, exactly similar to an object that is in motion in its space-time continuum and as you wear a physical body, there is an interaction between your human body and a Higgs field that is present in this inky void too as you are still in the universe (Just isolated from the rest world). Now,

A quantum mechanical interaction between the particles that make up your human body and this Higgs field, gives you mass everywhere and all the time that makes you something solid or a real physical stuff (or a true observer?). Something like this.

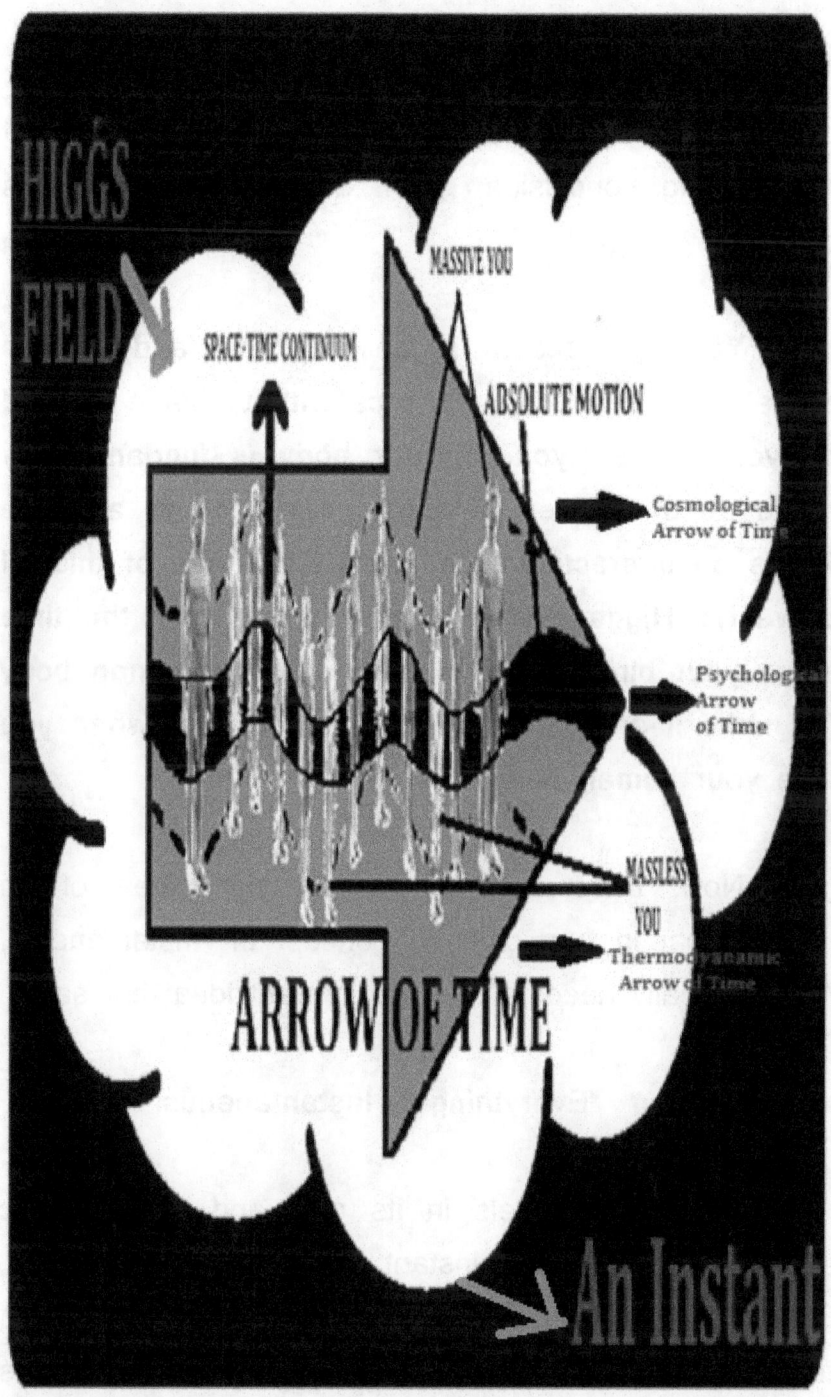

You don't think that there is something strange in this outcome? Well, I am sure you do but you are just busy to find out words to express your feelings about this starling conclusion. Yes? So, what is certain is this.

You are present in an inky void and as you are already in motion, perhaps without being noticed by you and as you or your body is fundamentally made up of smaller particles, your body as a whole keeps on interacting with an invisible sea of this all pervasive Higgs field, wherever you go, all the time since your birth when you are gifted a human body (to make it grow?) to the end of your life, when you lose your human body.

Now, there is a new idea, say "Idea of an instant" that in turn gives a concept of "Instantaneity". And we really need to use this brilliant idea that says

"Everything is Instantaneous."

Everything travels in its own and very specific space time continuum, instant by instant. For instance, Earth in its orbit around the Sun. [Because Earth maintains its motion with you, around the Sun in its own and very specific space time continuum, instant by instant what we call "Orbit of the Earth in the Solar System", quite similar to this.]

Earth's Orbit arround the Sun

Instant 1 [11:59:59]

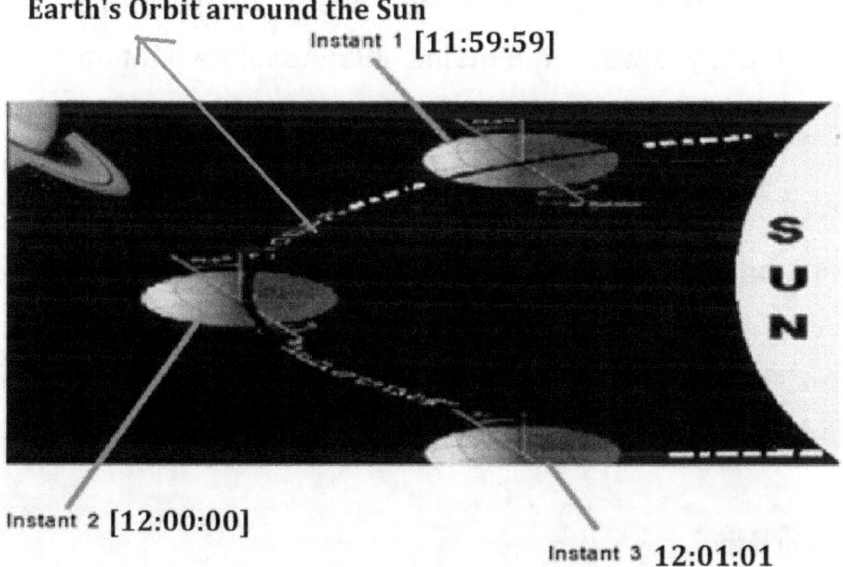

Instant 2 [12:00:00]

Instant 3 12:01:01

Now, the same holds true for any other physical like your own human body, with which you travel in the universe in your own and equally specific space time continuum, from one instant to another instant (What you call from your past to your future, through your present). So, in this inky void too, you are in motion and now our proposal is this.

Either you are in your imaginary inky void or you are on the floor of a metro train in your real world. Your body that is the human body you have,

always exists or travel in its own and very specific space-time continuum, instant by instant. And as your physical body is basically made up of particles like subatomic particles, so, every single particle in your body always maintains this Absolute Motion.

Now, you lose your massive physical human body in an instant and in next instant, when this instant becomes your past, you acquire it all back. And again and again, as your body's particles keep on interacting with this Higgs field and situation becomes something like this.

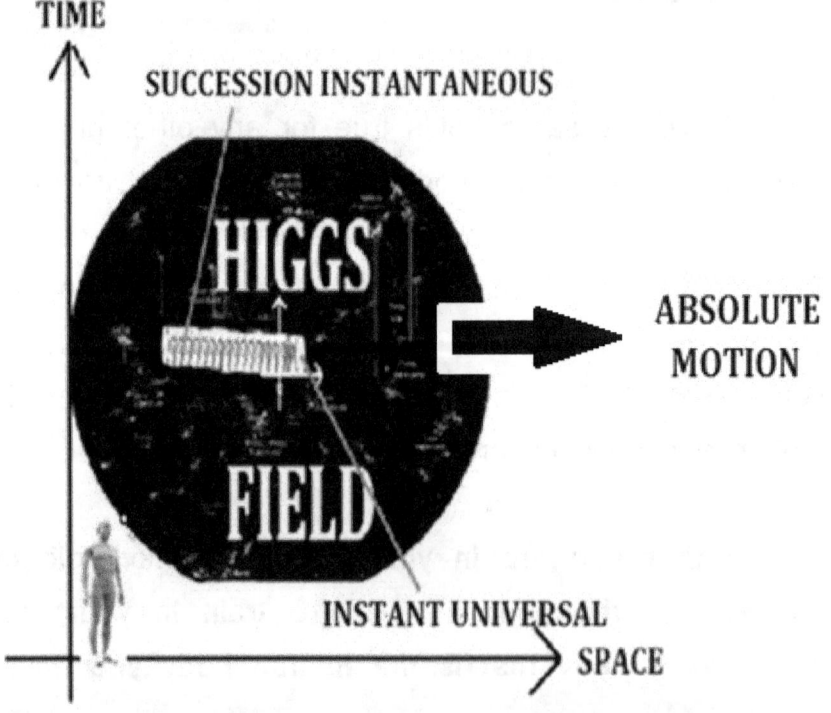

So, on one hand we have an idea of a sort of Absolute Motion that simply affirms the natural impossibility for the existence of true at rest (from the White Book) and on the other hand, we have an all pervasive Higgs field that gives mass to every single physical, everywhere in the universe, all the time (From the Green Book & Yellow Book).

An additional idea of instantaneous nature of a physical existence (From the Blue Book) also helps us to summarize our new hypothesis. Now, the hypothesis begins.

You are in your inky void, desperate to find something, so that you could develop an idea of relativity (That is the only reason I said this is an inky void as I am not satisfied with Einstenian idea of "Total Relativism"). You don't have any information that how big this void is and where it ends and whether it is already moving or standing still, somewhere in the universe.

[You might ask what "Total Relativism" is. Total Relativism is the concept that says the points of view have no absolute truth or validity.

They have only relative, subjective value according to the differences in perception and consideration.]

As we discussed, out of a sudden, you felt a jerk on your body and you said it is because of the fact that Galilean or Newtonian "Principle of Inertia" is valid everywhere in the nature, so, there was a change in the state of motion of this void, due to which I felt a force on my body in forward direction. Perfect.

Now, another possibility is that

As you are already in motion because if universe is expanding, your home galaxy Milky Way must be in Hubble flow or you can even leave all this stuff of expansion universal, etc if you find it confusing and just consider the fact that as we saw, if you take CMB as a background,

Earth with you and me is in a cosmic race, with a velocity of 390 km/s towards constellation "Leo" or simply presume that despite of the fact that no one ever told you this.

But as per the truth (or as per the cosmic or absolute truth) you have already boarded a space ship, since the instant when you entered in this universe or creation on your birth day, to approach

constellation "Leo" and the current velocity of your space ship is 390 km/s.

Second, we have an all pervasive field, we called as Higgs field and as it is ubiquitous, so, wherever you go or move, no matter you have enough senses or not to perceive or observe it, you are interacting with this field, all the time and everywhere in the universe.

Now, this interaction between your human body and Higgs field, creates an instantaneous nature of your existence, due to which in one instant you acquire a perfect mass for your body, in order to realize that you are a real physical stuff and in the next instant you lose it all.

And again in the next instant, as the particles of your body that are in motion with you, interacts with this Higgs field, as you are already in this sea, so, again you get your full mass back and in the next instant, you are again massless.

So, the true existence of yours in 4D Space-Time Continuum might look something like this.

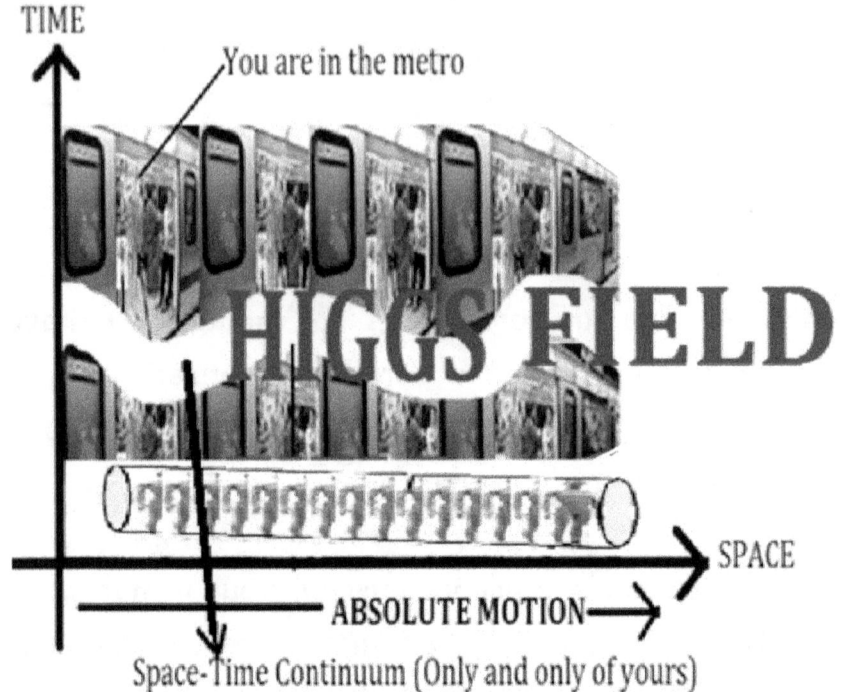

TIME

You are in the metro

HIGGS FIELD

SPACE

ABSOLUTE MOTION

Space-Time Continuum (Only and only of yours)

[Just to clarify this point, let me emphasize that the idea of "Interaction" like a physical interaction, itself is intrinsically "Instantaneous".]

Higgs mechanism implies that acquisition of mass by any physical must be "Instantaneous". So, if human body also acquires its mass from Higgs field, this phenomenon must be instantaneous too but it doesn't appear. Why?

See, any program that is executed on a television screen is executed in the form of packets. So, packets or quantum of the programs are displayed on the TV screen, one by one or instant by instant [as it is the case with a "Cathode Ray Tube" (CRT).] but the time interval between them is so small that when we watch the same program on TV, it appears as continuous. Similarly, human body like any other physical, acquires its mass from Higgs field, instant by instant but this interval between two instants is quite small (Moreover, as it is happening on quantum level, so, it must take time to travel to classical world here on earth surface), so, here we observe our human body, without any interruption continuously. Even whatever is existing in your surroundings is also in this creation flow, [and perhaps this is the way, power of The Matrix hides the truth from us because this absolute flow or creation flow just can't be observed by us or it exists beyond the limits of human senses.] But the duration of this instant is very small [may be even something like 10^{-43} seconds (Planck time)] or might be billions of billions of times smaller than a second and

It looks quite probable that a natural or cosmic instant can acquire any value between Planck time and the age of the universe that is the numerical value of an "Instant" can be anything between 10^{-43} seconds to 13.7 Billion Years.

So, here on earth surface you observe a human body all around you continuously.

Well, if you can see it. The real trick is going on somewhere else (at Quantum level or in Quantum world) and results are coming somewhere else (On classical level or on earth surface).

Just in case if you like "Magic Shows", this is exactly the same principle that a magician uses in order to show you his magic tricks and it is the only principle on which "Science of Magic" works. How? See, when a magician performs his magic trick, what you observe on the stage of the show is just an effect but the cause of it lies behind or even beneath the stage.

So, you are travelling in a very precise and specific Space-Time Continuum Wave, everywhere and every time and as a consequence you interact with this Higgs field that is the one and only source of your physical, massive and hence inertial existence that in turn gives you a human experience of what you call "Inertia".

It is this fundamental interaction that also gives you an instantaneous nature, which implies that your own true physical existence in the creation, travels in a space time continuum instant by instant (perhaps that is called life in general and your life in particular) and now if anything attempts to deviate you from this main course of flow like a physical force, it gives you a jerk or a feeling of a force, exactly similar to a classical force, what you just felt in your inky void or on the floor of the metro train.

So, you as well as I can see now that a natural happening, that is known as "Inertia", might still have a more mysterious aspect then it appears and this "Law of Inertia" that is the starting point of Newtonian or what is called as Classical Mechanics in the science curriculum throughout the globe today is not what it looks like and its simple appearance might just be Platonic "Power of Appearance".

Because in relation to you as an observer, it's obvious that there are coordinate systems or what we said frames of reference (or even worlds) within frame of reference, within frame of reference .....Ad infinitum. Or

"Worlds within the Worlds".

Something like this.

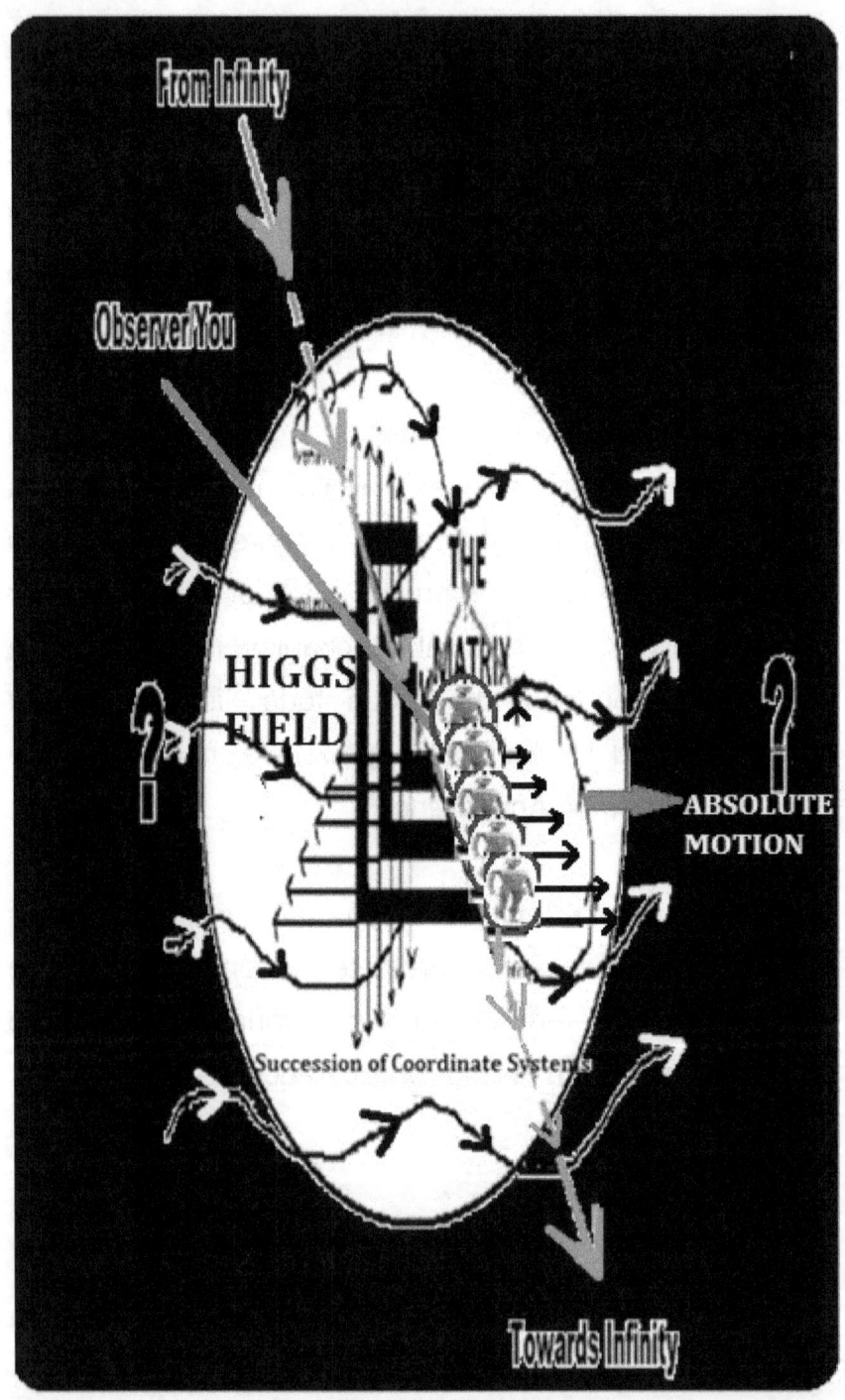

That is perfectly analogous to our puzzle of "Box within the Box" or "Particle within the Particle". And hence there could be a chaos of appearance within the appearance, within the appearance, within appearance ....ad infinitum. Perhaps this is the way how "Power of appearance" operates in the universe.

And the final conclusion of this whole is that the truth of inertia is still far more mysterious and the science of Inertia remains unknown still. And if it is the case, it will give disastrous consequences to the whole body of modern science. Because if it is true than the next obvious conclusion that can be drawn is this.

There is something wrong with the "Modern Science", at its very fundamental level.

[T]

Modern science or perhaps we should say the most advanced research and technology of the day, on the subject, says that if you go deep inside,

"There is no mass in any physical stuff."

But it is the vibrations in the subatomic realm that causes the illusion of mass for us. But I doubt that because if it is the case, there has to be something that vibrates at its very first place, in order to create a cause for the effect that is solidity inside of any physical object or existence but they say that there is none because there is no mass at all.

Having said that, we must reconsider our final conclusion, what we got from "Yellow Book" that the ultimate source of inertness or solidity, what we say its inertia is basically coming from an absolute motion that is an intrinsic truth or a fundamental property of every single physical. So, the reason of solidity or inertness in any physical existence is not the vibrations at its quantum level but a sort of motion that has never been predicted or realized in the history of modern science till date. And this is the same thing what we are saying as "Absolute Motion".

Now, as we have observed in the "White Book" that in the entire history of science, perhaps the most ignored concept or science is the "Concept or Science of Observer". But as we have seen so far that the most important thing in this whole journey is "Observer" itself. Because it is the observer, who is the ultimate domain of this "Illusion of Solidity" itself and as we saw in "Yellow Book" in terms of "The Trinity of 3 Os" viz "Observation" (Nature), "Observable" (Cosmos) and "Observer" (You) that the most important thing in this absolute trinity (or "The Trinity") is "Observer" that is you. So, it is quite obvious that in order to develop true mechanics of this cosmos, we need to reconsider the importance of the concept or science of observer. Perhaps it could be the realization of all those Newtonian giants too but unfortunately it didn't. Besides, if you really understand the "Theory of Relativity" (Special and General), you can see that it is basically an inherent idea of "An Observer" that makes the theory itself.

There is no such thing in modern science today but I would like to call it as "Observergenesis", a science of the synthesis or more precisely the genesis of an observer of this creation like you (or me or any other human being). Now, this term Observergenesis (Observer Genesis) is basically an idea that how an observer comes and evolves in this universe, in order to do its prime function that is the "Observation of this Grand Design or this real world" [Or this Platonic "Power of Appearance"?] because we have advanced technologies today like "Ultrasound", "MRI" (Magnetic Resonance Imaging) etc. So, we are capable of finding out the truth of this "Observergenesis". So, here it begins.

Your mom and dad, a final stroke and a union that happens in this macroscopic world or at this classical level or on earth surface that gives a quantum entity known as "Sperm". And a run starts, in which your father's sperms, without having any intention to give you your birth, starts running towards its prime destination that is an "Ovum" in your mother's womb, which is similarly a quantum entity. And it might look like this.

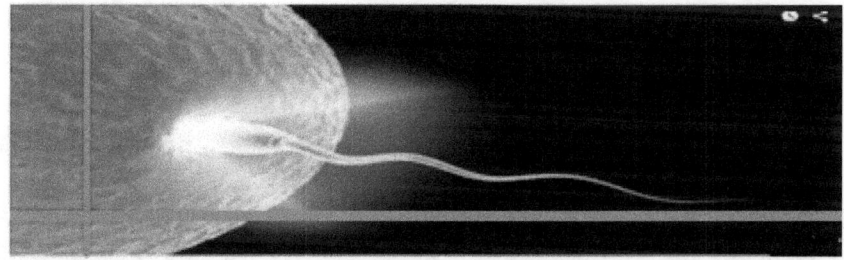

[Well, if you see it deeply, it appears that the true cause of your birth through "Observergenesis" starts when your dad makes love to your mom and this instant or event is happening at macroscopic or classical level on Earth surface but the genesis of sperm and ovum is from molecular arrangements and hence its real cause is microscopic/quantum world. But the fusion of sperm and ovum is an event that happens somewhere between macroscopic or classical world or Earth surface and microscopic or quantum world.] And now we move on with this unification (Yoga) of a sperm and an ovum. As you already know, the end result of this is a "Zygote", perhaps an absolute domain of life in general and your life in particular.

At this point of observergenesis, what happens next is cell divisions, millions of cell divisions and this is called "Organogenesis", where there is continuum of your body's manufacturing (that surprisingly look exactly similar to "3D Printing" of modern day technology), in terms of synthesis of its different parts or organs like "Eyes", "Ears" etc. and when the absolute domain of your human senses like "Vision" or "Touch" etc also gets synthesized. [Please note that the true origin of human senses is from this point.] And from here on, an observer or you (or me or any other human) are going to have your different parts and if you like, you can notice that this whole natural processes is a perfect match for the idea of "Pattern Formation" of chaology (Science of Chaos). Because it is simply fascinating to observe here that an otherwise a perfect molecular chaos inside a human body becomes the ultimate creator of a new creation that is your physical body, all around you. And not only is this but within the body of this transformation of chaos into creation, there is an inherent idea of "Control". But it's a bit complicated because who has this control?

And who is under control?

So, again you can see that it is this chaos, who is the ultimate creator of the creation. And creation is of an animate object or a living being or it is of an inert or in-animate object doesn't really matter. But what is amazing here in this development of an observer or you are the sequence of events and there is something really noticeable hare. Let's see what it is.

As we discussed, only two cells are needed in order to start a process of what we called "Observergenesis" in general and your birth in particular, only two cells, a sperm and an ovum (And a miracle?)

And now, the unification of sperm and ovum happens inside your mother's womb and as a result you enter in this universe but just isolated from the outer physical world. Something like this.

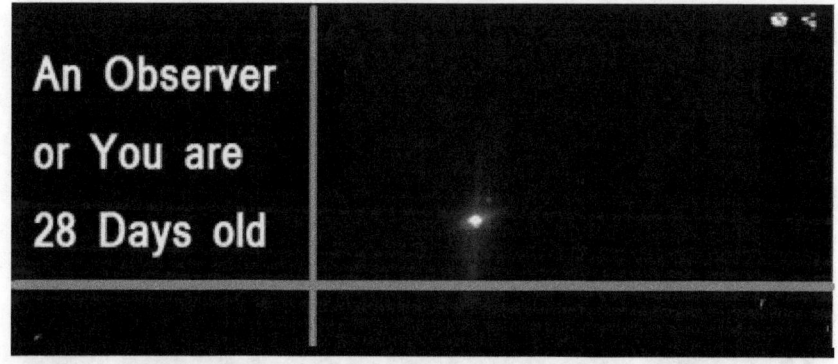

An Observer or You are 28 Days old

And now,

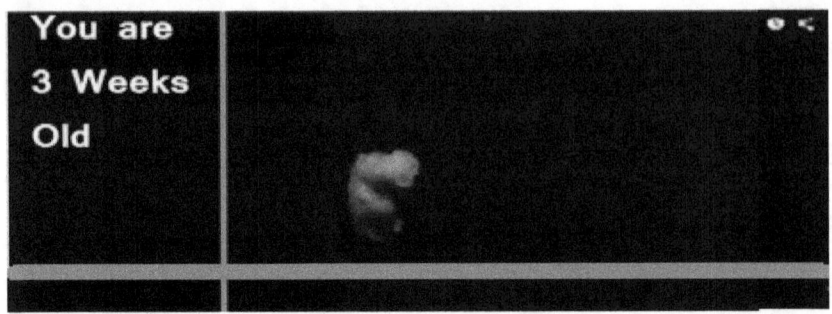

And now,

And now,

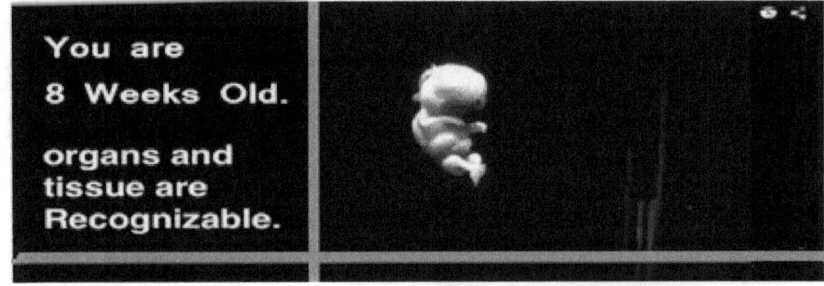

And now,

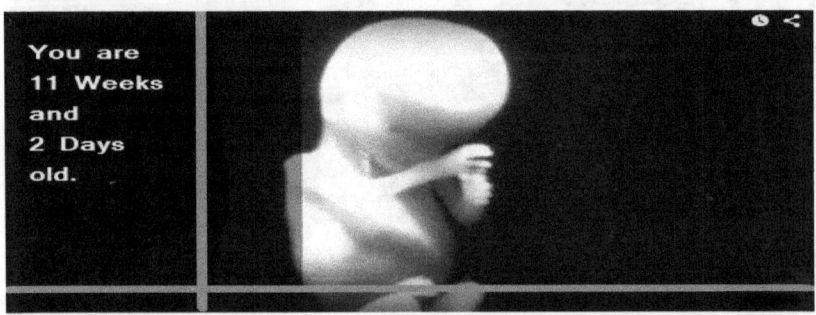

And now,

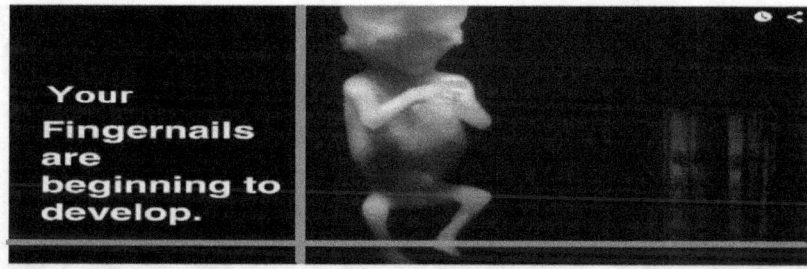

And now,

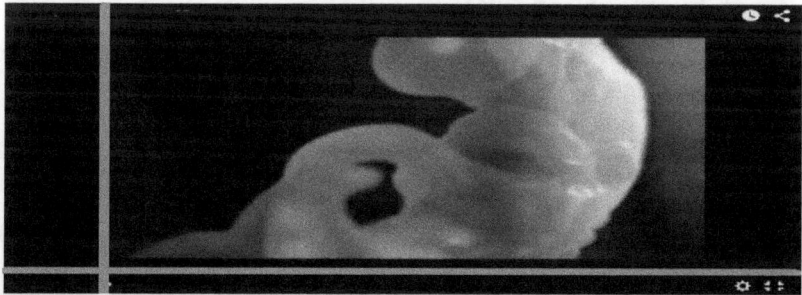

You got a heart (But not emotions?). *[Look at this, as you know, whole of the modern research is based on the presumption that the human brain is the prime thing in our body. But here it is obvious that it is human heart that is the main thing and later on human brain comes in this process, like an assistant.]*
And now,

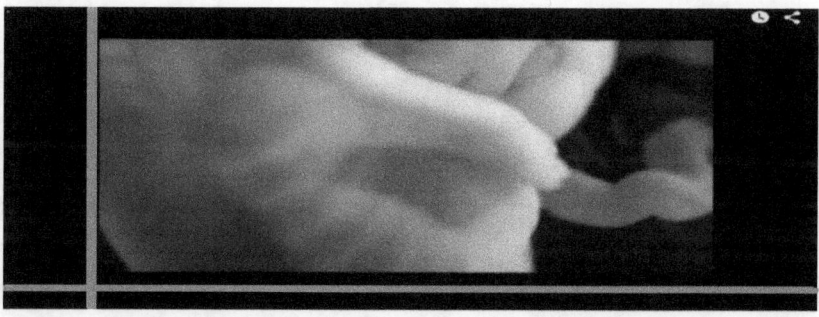

You got two hands and now,

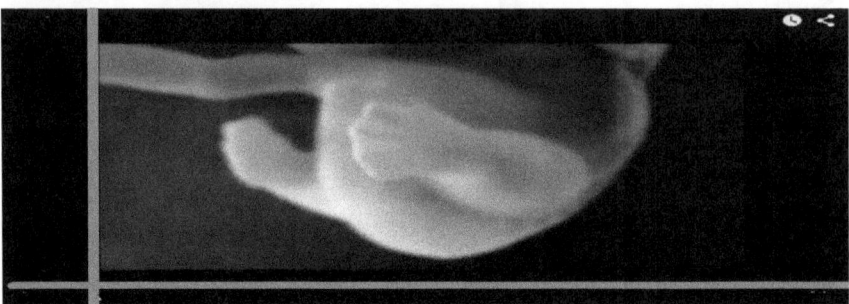

You got two legs and now,

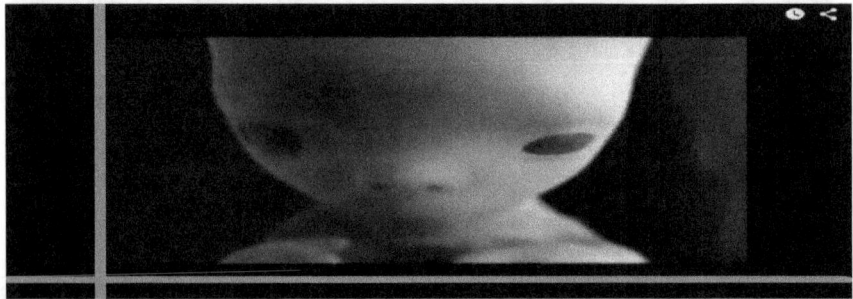

You got your eyes (But not vision?). Now,

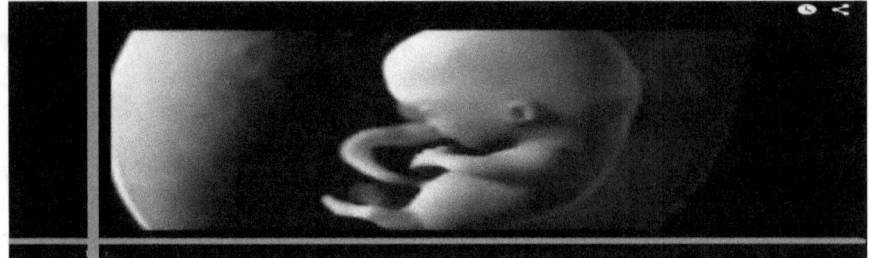

You got your ears (But not hearing?) and now,

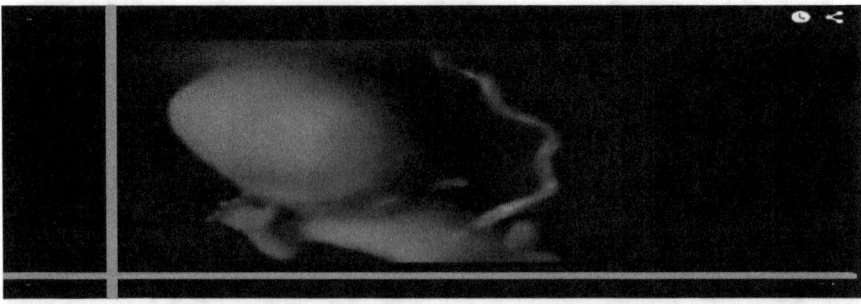

You got your brain and hence an "Operator" to control your body. Now,

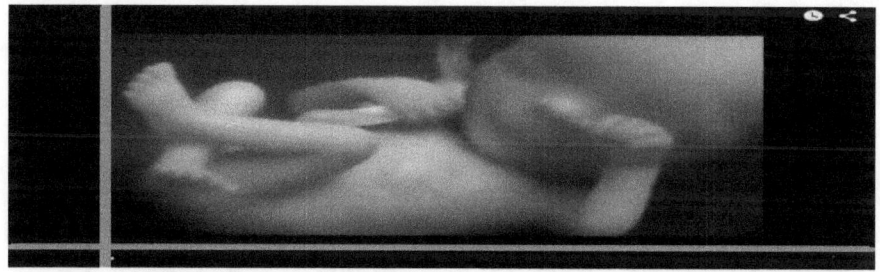

You are ready to leave your mother's body. And it is this point on which you are about to break the isolation of yours from the rest physical world. And here you go.

Welcome to the wonder world. Finally here you are in the hands of your father. Perhaps the very first observation of yours, as a new born or as a natural observer is the touch of this world by your touch senses or through your skin and because this experience is entirely new for you, so, what you do in the next instant, you start crying. Because of the fear of an unknown and this is nothing but a survival instinct.

Now, it is for sure that the mode by which "Observervergenesis" happens and as an outcome, an observer like you comes into existence, there must be something that we can call "Observotermination", when an observer like you, goes out of existence. And the mode by which an "Observotermination" (Observer Termination), when an observer like you dies and the mode of "Observervergenesis", these two apparently different modes must be somehow related to each other. And the mode of "Observervotermination" or your death is exactly similar to its counterpart "Observervergenesis" but exactly in reverse direction. So, now we need to see how an observer or you go out of existence. In other words what happens when this most wonderful creation of the universe or our creator that is this beautiful human body, dies?

Death is as natural as birth and at the moment of death the human brain has a surge of activities as it gets less and less amount of oxygen and then it goes completely dark. Although some process remain continuous, like heart beating but most of the other processes in the body stop and what is called as

"Post Death Process" begins.

First, all your muscles relax. "Algor Mortis" or "The Chill of Death" begins immediately at death, so, your human body starts cooling at the rate of 1.5 degrees per hour, until it reaches to the room temperature. Eventually your surroundings start eating your body and you are gone forever (Dust to dust and ashes to ashes?).

And if you would like to know in detail about this "Observotermination" then there are few stages of the process. First step is "Initial Decay". In which your body cells die and the bacteria that lived with you forever, starts proliferating, as they got a life-time achievement award and it might look something like this to us.

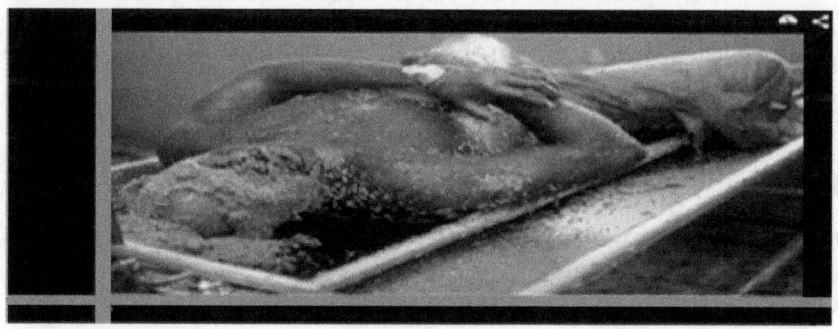

Now, in the second stage called as "Blot", gases produced by those bacteria get accumulated and your body smells horrible and as an example it might look like this.

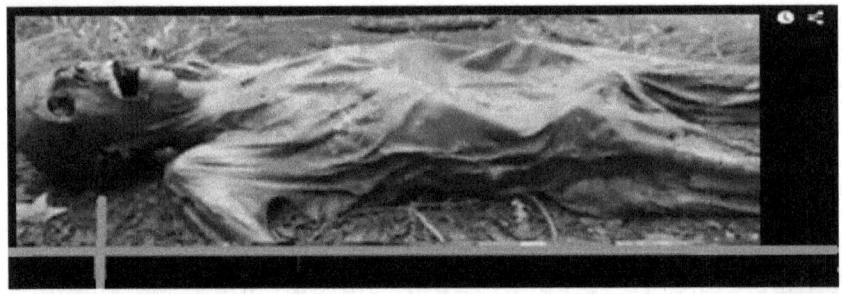

Finally, the third stage, when your body releases liquids and gases and skin gets ruptured. Now,

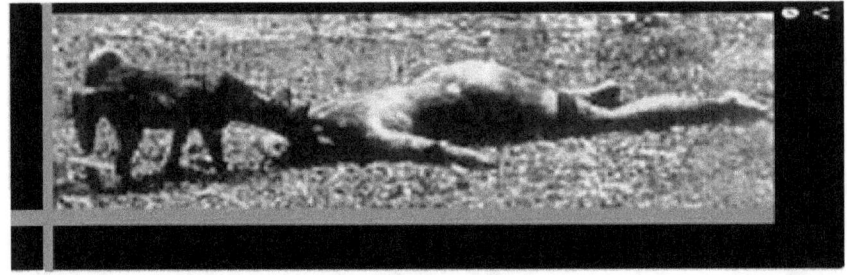

And in the next stage, your lovely human body gets disappeared in dust.

May be we can talk more about the fate of your body after you are gone from this world. As we have seen that your human body after your death goes through many stages, until it finally gets disappear into its ultimate source. First, heart stops beating then skin turns a light gray, then all your muscles relax. Now, the bladders and bowls get empty. Then, your body temperature drops down. Then your skin gets purple and waxy. Then your lips, fingers and fingernails get fade to a pale color or white as the blood leaves. Then, blood pools at the lowest parts of your body, leaving a dark purple-black stain called as "Lividity". Then your hands and feet turn into blue. Now, your eyes start sinking into the skull that might look like this.

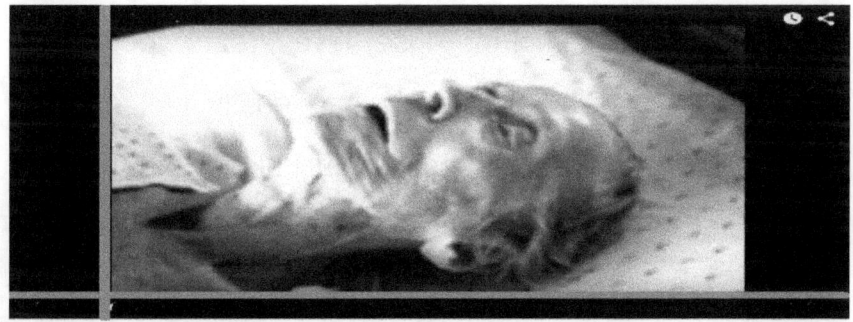

And then rigor mortis or relaxation of your muscles continues and purpling of the skin and pooling of the blood also continues. Now, rigor mortis begins to tighten your muscles for about another 24 hours, that makes your body tense and then it gets reversed and the body will return to a limp state.

After 2 hours from your death moment, your body is in full rigor mortis. After 24 hours, your body temperature gets equal to the surrounding environment. Now, in males semen dies, your neck and head are in a greenish blue color, this greenish blue color continues to spread throughout your body. Then comes

a strong smell of rotting meat and your face, the best tool of your identity in this real world is no more recognizable. After 3 days the gases in the body tissues forms large blisters on your skin. The whole body begins to swell and bloat. Fluids inside your body try to leave you, through any way they might get, like through your mouth, nose, eyes, ears, penis and rectum. After 3 days, anybody can easily pull off your fingernails from your corpse. Your skin gets cracked and burst on many places, mainly because of the internal pressure and because it itself tends to get decomposed. And this decomposition remains continuous, till the time when your body is nothing but the remains of your skeleton. Like this.

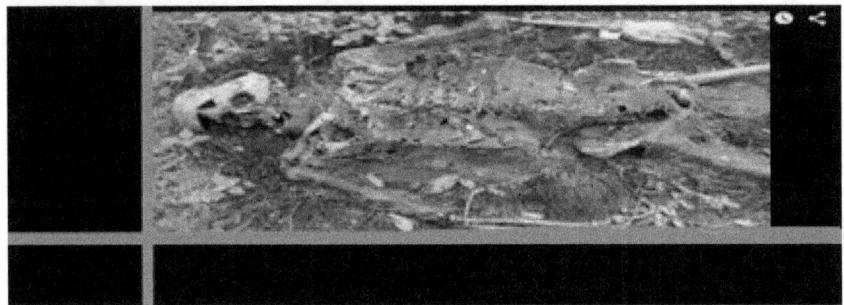

Well, it is quite confusing because where we were? And now where we are? But I do have a point. See, starting from "Green Book" including "Yellow Book" and "Blue Book", it is damn obvious to us that if mass acquisition by any physical is through an all pervasive quantum field, what we called "Higgs Field" then all of these natural phenomenon like "Observergenesis" (Observer Genesis) and "Observotermination" (Observer Termination) must have an intrinsic link with this Higgs field and this must make all these apparently isolated events or instants, "Instantaneous". So, starting from the first instant when your mom and dad come together till the instant when your body gets dissolved into dust, this whole stuff must be happening in its unique space-time continuum, instant by instant, in this mysterious Higgs field. So, the same idea of "Instantaneity" that was true when you were in your imaginary inky void or when you were in the metro, in your real world, must be true here as well, when you are in your mother's womb and when you are in a coffin. *[Moreover, as you can see, this is a perfect point for the grand unification of modern science and spirituality or even a unification of what seems "Scientific" with what appears "Magical". But as of for now, just leave it.]*

[U]

n order to find out the absolute origin of "Power of Appearance" of this Universe or in this universe that in turn might be helpful in order to get our new understandings about "The Illusion of Solidity" and "Inertia", we must reload the sole idea of Plato's "Allegory of The Cave", [The Republic, 514a- 520a] when he told us about the "Shadows in the Cave" and about the "Power of Appearance", as we saw it before that Plato (428/427 or 424/423 BCE, Athens) said in his own words,

"Power of appearance, led us astray and through us in confusion. Whereas,

Art of measurement would have caused the soul to live in peace and quiet abiding in the truth."

As a matter of fact, I do have an elegant example for you, to prove this point. Look, in order to discover this proof, what all you need is a device or a thing, in which there is a wooden handle and at the middle of this handle, there is a wooden circle. So, this arrangement might appear like this.

Now, on one face of this wooden circle, there is a drawing of a "Parrot" and on another face of this circle, there is a drawing of a "Cage". Something like this.

[It is the first face of the wooden circle, with a painting of a Parrot.]

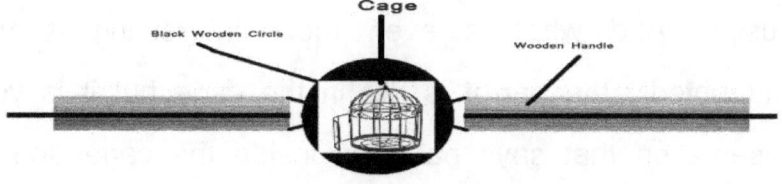

[It is the opposite face of the same wooden circle, with a painting of a Cage.]

Everything is fine. Just because there is no motion and this arrangement is on "At Rest". Now, you insert an idea of motion in this arrangement or say you start spinning this wooden handle.

What happens? Undoubtedly, if you are spinning this handle fast enough then for you as an observer, it will look as the parrot is inside of the cage.

Something like this.

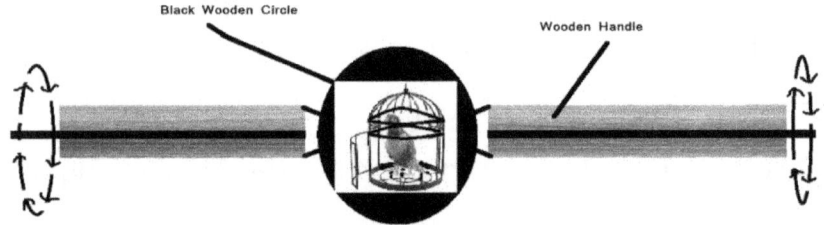

So, you can see. Parrot is truly nowhere in the cage. But when these two isolated appearances, namely "A Parrot and "A Cage", unify in their true domain that is motion (that in this case is spinning), they create another appearance or a third appearance "A parrot inside a cage" that simply is a perfect illusion. And what is even more interesting is that undoubtedly the parrot is not in the cage but it is your observation that says parrot is inside the cage and all the human emotions such as your sympathy for this poor bird, because now it is apparent to you that this parrot is not free to fly in the sky and your mercy for it, come along with this natural observation spontaneously that is far from being true. Now, the question is how is this possible? Or how it happens? Well, Of course you can say that this example is mere a kid's play. Nevertheless, this simple demo is quite a raw stuff, in order to see, how this universe originate appearances (what we have seen in "Blue Book" in terms of our conundrum of "Appearance within the Appearance") and how a real physical existence, do exists in this nature, at its first place. At least now you can realize that the appearances get unified into their true domain that is "Motion". And this is how "Power of Appearance" operates in this universe. Because the true domain of inert or inertness or inertia and mass or solidity is motion. So, these two apparently isolated natural operations at a certain level in the whole interval of space, from 10^{-35} m (Planck Length or quantum world) to 45.7 Billion light years (Cosmic Realm) via 1 m level or through this classical world on earth surface and this whole interval of time from 10^{-44} seconds (Planck Time) to approximately 13.7 Billion Years (age of our universe) via 1 second on earth surface, where there is an origin of an appearance and origin of solidity in this real world are basically one and same things or are simply two faces of a single coin and hence the absolute source of solidity as we saw earlier in "Blue Book" that is absolute motion, is in fact the true source for the origin of appearance in the nature too. Unbelievable? Okay.

You know what. I must confess that the real hint about this outstanding idea came to me from the same mysterious ocean of ancient text, when it told me that

"For the origin of a physical existence, there is a need of nothing but "Nimitta" (Cause) because physical or natural existence gets its "Existenceness" from its own expense or its geometry itself."

And again, this doesn't matters that this "Thing" or "Existence" is of an animate or a dynamic stuff like you and me or it is of an inanimate or inert stuff like a stone. Moreover, combined with the idea of "Pattern Formation" of "Chaos", this synthesis of appearances in this natural world, in this space-time interval of 10^{-35} m to 45 Billion light years & of 10^{-44} seconds to 13.7 Billion Years can give you an exact picture about another cosmic or natural continuum called as "Change" or "Cosmic Change".

Because in this whole interval of Space-Time,

the appearance is of a "Dinosaur" or it is of a human being, doesn't matter at all.

And again, combined with the idea of "Instantaneity" as we saw in "Blue Book", we can see that the whole continuum of synthesis of an appearance in its unique and specific space-time continuum is fundamentally "Instantaneous". It is obvious that the appearance of a "Dinosaur" was instant by instant, in their space-time continuum as it is the case with an observer of today like you, where we have already seen its true nature that is "Instantaneous". Although it looks like a perfect natural continuum, just because every single appearance is bounded by the laws of nature, so that it always maintains its own space-time continuum, instant by instant because it remains in its true domain that is motion or "Absolute Motion".

And now the time has come when we can create a rainbow, by mixing the colors of our all 4 books viz "White Book", "Green Book", "Yellow Book" and "Blue Book".

[CONCLUSION]

Modern science on the subject says that there is a power in the universe, this power lies within the matter that exists in its domain that is a physical or a natural object or an existence. It's this power that enables every physical to resist any change in its present state of existence (or state of motion, to be precise).

They say that this power is "Inertia", a sort of inertness and source of this power is within every physical existence. But the source of this power of a physical existence is somewhere else in the universe.

Whereas, the Power of "The Matrix" is concentrated within a single object and the object is "Observer", that is "You". So,

Whatever you observe in this nature, you observe it, because and only because you are allowed to observe it.

As if the human senses like vision or touch, etc (Gifts of God to humankind?) themselves are nothing but the "Absolute Limits".

Even more unfortunate for humanity is the "Axiomatic Origin of Modern Science" from ancient times, like Pythagoras (570-495 BC), Euclid and Aristotle. Because origin of an "Axiom" itself is from the natural observations and the creator of every natural event is "The Matrix" itself. So, there is no way you or they could take it for granted that an axiom is a fact of nature or a cosmic truth.

Alas, they all did the same anyway and as we saw in this book, there is a strong probability that says that they could never defeat the "Power of Appearance".

The Matrix is so powerful in this universe that without even applying any real trick, within an interval of 90 Billion Light Years (expense of the Observable Universe) to 10^{-32} cm (Planck Length, far smaller than an atom) in space and in the interval of 13.7 Billion years (Age of the Universe) to 10^{-44} seconds (or Planck Time) in time, it creates,

Particles within the particles

&

Structures within the structures

&

Chaos within the chaos

& even,

Worlds within the worlds.

In the beginning of this book, I asked you to imagine an inky void that is perfectly empty and your existence inside of it. But if you really did the same, I don't think now it's hard for you to observe a great similarity between this inky void and your everyday world. As this world is mostly a void, just not so empty as your imaginary void and even better, quite similar to your imaginary void, you don't know its expense either or do you? Suppose, you wake up in a dark room one day and you switch on a light bulb and you realize that the room you were previously thinking empty is in fact, full of stuff around you, the only difference is that time it was not just observable or a subject matter of your human senses. Quite similar to the situation,

"Let there be light ....&..........There was light".

And when there was light or let's say "Real

Light", whatever was unobservable or unknown before or whatever was beyond your human senses became observable or known or the subject matter of your human senses. Well, you don't need to be religious for this but this is exactly the idea or the fundamental principle of "Yoga".

You wake up every day in a void. The real matter is not the fact that how much stuff it contains or how much it is empty but the fact that how much do you know about it.

[END?]

[No, trust me on this, it's a beginning.]

[Let me give you a hint. If you take this brilliant idea of "Instant" & "Field" and their intrinsic link with you as a cosmic existence, at its face value, as we did in this book. It is for sure that you can write a thesis on it. As much as you think about it that much it will expand. Why don't you just give it a try and see for yourself.]

fulltonic@gmail.com

www.ingramcontent.com/pod-product-compliance
Lightning Source LLC
Chambersburg PA
CBHW051857170526
45168CB00001B/135